# Geology of Cumbria

## Lakeland's Rocks and Minerals Explained

by

### R. V. Davis

DALESMAN BOOKS

1977

80p.

**THE DALESMAN PUBLISHING COMPANY LTD.,
CLAPHAM (via Lancaster), NORTH YORKSHIRE**

**First Published 1977**

**○** R. V. Davis, 1977

ISBN : O 85206 407 1

Grateful acknowledgement is extended to Pauline Rogers for help in the preparation of the manuscript.

**The back cover photograph shows crystals of the rare lead mineral Campylite. These orange-coloured, barrel-shaped crystals are found on the Caldbeck Fells.**

*To Rosemary and Amy*

Printed in Great Britain by
GEO. TODD & SON,
Marlborough Street, Whitehaven.

# Contents

The work is set out as a series of questions and answers :

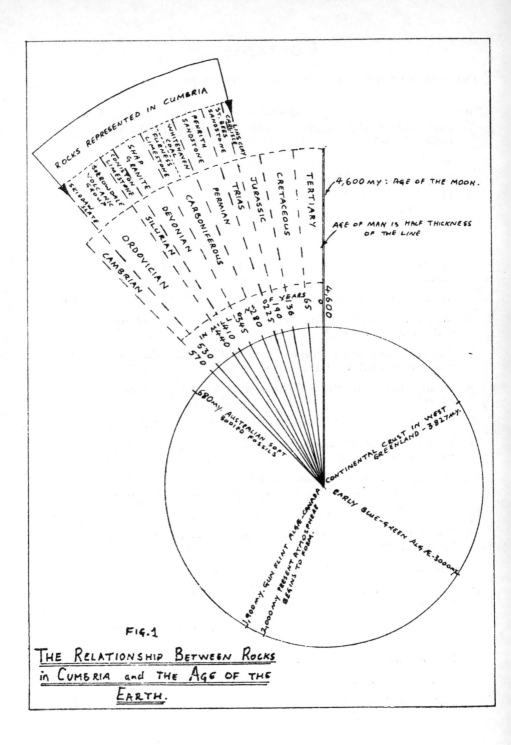

ROCKS REPRESENTED IN CUMBRIA

SKIDDAW SLATE
BORROWDALE VOLCANIC GROUP
CONISTON LIMESTONE
SHAP GRANITE
CONISTON LIMESTONE
COAL
WHITEHAVEN LIMESTONE
PENRITH SANDSTONE
ST. BEES SANDSTONE
CARLISLE
LIAS CLAY

CAMBRIAN
ORDOVICIAN
SILURIAN
DEVONIAN
CARBONIFEROUS
PERMIAN
TRIAS
JURASSIC
CRETACEOUS
TERTIARY

MILLIONS OF YEARS
570
530
440
410
345
280
225
190
136
65
0

4,600

4,600 MY : AGE OF THE MOON.

AGE OF MAN IS HALF THICKNESS OF THE LINE

680 MY. AUSTRALIAN SOFT BODIED FOSSILS

CONTINENTAL CRUST IN WEST GREENLAND - 3,827 M.Y.

EARLY BLUE-GREEN ALGAE - 3,000 M.Y.

1,000 M.Y. PRESENT ATMOSPHERE BEGINS TO FORM.

1,800 M.Y. GUN FLINT ALGAE CANADA

FIG. 1

THE RELATIONSHIP BETWEEN ROCKS in CUMBRIA and THE AGE OF THE EARTH.

# GEOLOGY OF CUMBRIA

1. **How wide is the variety of rocks found on the Earth's surface? What types can be found in Cumbria?**

There is an almost infinite variety of rocks to be found on Earth. Rock names are convenient classifications for groups of minerals held together in definite patterns and proportions. Therefore a slight change in one mineral can cause the rock to have a different name. To generalise we use a simple three-fold classification:

**Igneous** — fire formed.

**Sedimentary** — usually water formed.

**Metamorphic** — existing rocks altered by heat and pressure.

In Cumbria we have **all** of these rock types present. At Shap we have three quarries within a few miles of each other — Shap Pink Quarry contains **Igneous Rock** (Granite); Shap Blue Quarry contains **Metamorphic Rock** (Hornfels); and Shap Limestone Quarry contains **Sedimentary Rock.**

2. **What is the significance of having all three rock types present in such a small area?**

i) Where **igneous** rocks are located, they represent a point where the crust of the Earth has been weakened at some time in the geological past. [1] The central mountainous part of the Lake District is formed largely from igneous rocks.

ii) Evidence suggests that some of the oldest exposed rocks in our area are **Sedimentary** in origin. This implies that there has been an alternating series of sea-time and land-time for many hundreds of millions of years. The nature, content and position of these sedimentary rocks when compared with the igneous rocks can tell us about conditions in the past [2].

(1) The appearance of igneous rocks is not uniform—they have different colours and texture depending on how they were formed and what they contain.

(2) Shale beds in Carboniferous Limestones indicate periods of slight earth uplift during an otherwise calm period.

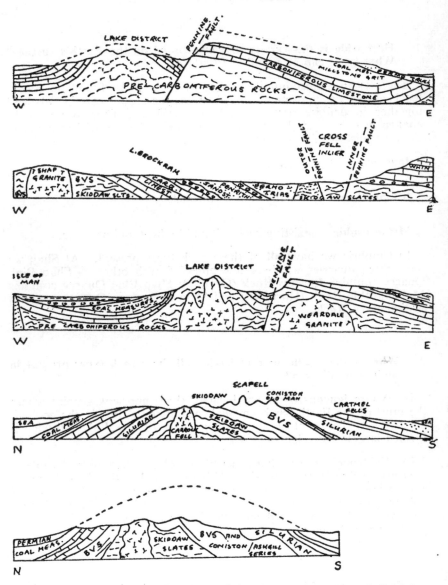

Fig. 2 : Selection of typical pre-plate theory cross-sections of Cumbria.

iii)  The great heat generated by the igneous rocks whilst they were cooling down from a molten state, would have had some effect on the surrounding rocks.  Rocks which are altered in some way by such action are called **metamorphic** rocks, and any existing rock can be metamorphosed.

### 3.  How do local rocks fit into the overall picture of the age of the Earth?

The oldest rocks yet dated are in the order of 4,600,000,000 years old,[3] and geological sciences are trying to piece together the developments which have taken place ever since.  Our present knowledge is very limited and currently geology is going through a period of "Conceptual Revolution".  Think, for example, about Plate Tectonics and Terrestial Magnetism — we are rejecting many "sacred" beliefs and casting around for something to take their place.

The rocks in the British Isles date from Pre-Cambrian times and we are mainly limited to the last quarter of Earth history; this is the last 600 million years or so.  The study of the history of Earth development is called **Stratigraphy,** and the more immediate past is divided up into periods of time which reflect a particular happening rather than just a precise unit of time.  The Stratigraphic Periods which are represented in local rocks are Ordovician, Silurian, Devonian, Carboniferous, Permian, Triassic, Jurassic[4] and Recent. These are shown as part of the 24 hour clock of the Earth's history in Diagram 2.

### 4.  Is there anything we can discover from Cumbria rocks which ties in with any new Geological ideas?

Most geologists tend to agree that the area presently covered by the Lake District has been a land mass at times when much of the rest of the country was underwater, although we must remember that we have also been submerged for vast periods.   The central core of the Lake District massif is of a granite type rock.  Geological cross-sections from some of the earliest text-books have shown this granite mass underlying the Lake District.  Some of these are shown on the opposite page.

What the books did not explain was the reason for why the granite is there.  The simple view was that a mass of molten lava had forced itself into a natural "oven" within the Earth's crust.  Here it had cooled down to form igneous rocks with a characteristic large crystal-line structure, so indicative of a slow cooling rate.  This idea did not explain why the granite had chosen to intrude where we are.

(3)  The oldest rocks known at the moment occur in Central Africa and Greenland.

(4)  Jurassic rocks do not outcrop on the surface, but they have been proved under the glacial drift near Carlisle.

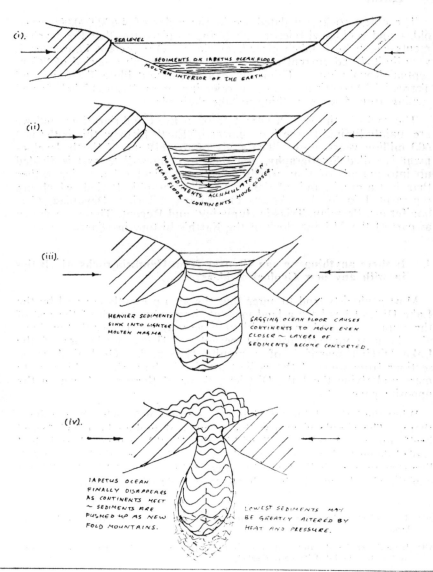

OLDER IDEA OF SAGGING OCEAN FLOOR PULLING TOGETHER TWO CONTINENTS

(i).

SEA LEVEL

SEDIMENTS ON IAPETUS OCEAN FLOOR

MOLTEN INTERIOR OF THE EARTH

(ii).

MORE SEDIMENTS ACCUMULATE ON
OCEAN FLOOR ~ CONTINENTS MOVE CLOSER

(iii).

HEAVIER SEDIMENTS
SINK INTO LIGHTER
MOLTEN MAGMA

SAGGING OCEAN FLOOR CAUSES
CONTINENTS TO MOVE EVEN
CLOSER ~ LAYERS OF
SEDIMENTS BECOME CONTORTED.

(iv).

IAPETUS OCEAN
FINALLY DISAPPEARS
AS CONTINENTS MEET
~ SEDIMENTS ARE
PUSHED UP AS NEW
FOLD MOUNTAINS.

LOWEST SEDIMENTS MAY
BE GREATLY ALTERED BY
HEAT AND PRESSURE.

**Fig. 3a**

NEWER IDEA OF OCEAN FLOOR SPREADING AND PLATE MOVEMENTS.

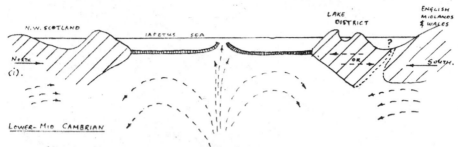

(i).

N.W. SCOTLAND    IAPETUS    SEA    LAKE DISTRICT    ENGLISH MIDLANDS & WALES

North    OR    South.

Lower- Mid CAMBRIAN

CONVECTION CURRENTS IN MOLTEN EARTH'S INTERIOR HELP TO CAUSE SEA-FLOOR SPREADING AND CONTINENTAL DRIFT. LAKE DISTRICT COULD HAVE BEEN AN ISLAND ARC WITH A SPREADING SEA FLOOR BETWEEN LAKE DISTRICT AND WALES ALSO.

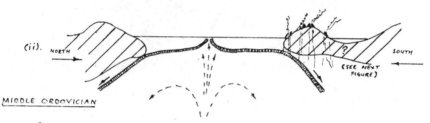

(ii). NORTH    SOUTH

(SEE NEXT FIGURE)

MIDDLE ORDOVICIAN

SUBDUCTION ZONES OCCUR AT THE CONTINENTAL MARGINS AS SEA-FLOOR ROCKS ARE FORCED DOWN. BORROWDALE VOLCANICS ERUPTED.

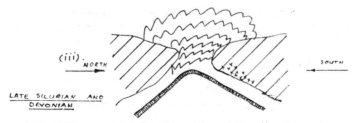

(iii). NORTH    SOUTH

LATE SILURIAN AND DEVONIAN

CONTINENTS COLLIDE; SUBDUCTION COMES TO AN END; IAPETUS SEA SEDIMENTS ARE FORMED INTO NEW FOLD MOUNTAINS; GRANITES HAD BEGUN TO FORM AT DEPTH.

**Fig. 3b**

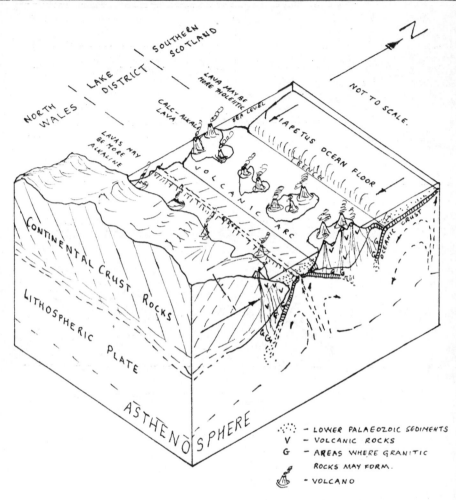

NOT TO SCALE.

Legend:

⋯ = LOWER PALAEOZOIC SEDIMENTS
V = VOLCANIC ROCKS
G = AREAS WHERE GRANITIC ROCKS MAY FORM.
🌋 = VOLCANO

FOR FULL EXPLANATION OF STRUCTURE SEE — WRIGHT, A.E., 'ALTERNATING SUBDUCTION DIRECTION AND THE EVOLUTION OF THE ATLANTIC CALEDONIDES' — IN NATURE Vol. 264 Nov. 11th 1976 Pg 8, Fig. 1.

FOR FULL EXPLANATION OF VOLCANISM SEE — STILLMAN, C.J., 'PALAEOZOIC VOLCANISM IN GREAT BRITAIN AND IRELAND'. — IN Jl. GEOL. SOC. LONDON. Vol. 133 1977, pp. 401-411. (THIS CONTAINS A REPORT OF THE VOLCANIC STUDIES GROUP MEETING HELD AT TRINITY COLLEGE, DUBLIN, 24th. SEPTEMBER 1976.)

# PLATE TECTONIC EXPLANATION FOR THE FORMATION OF CUMBRIA.

**Fig. 4**

The recent theories associated with **plate tectonics** have enabled geologists to put together many of the individual pieces of knowledge [5]. If a plate of land moves on the Earth's surface, then two contrasting situations can arise — (i) where two plates **move apart** from each other then we might expect to find the **ocean floor spreading**; (ii) where two plates **move towards** each other then we might expect to find a geosyncline being compressed so that the sediments are contorted and uplifted into new fold-mountain [6]. Of course, these two alternatives are not mutually exclusive and an infinite number of other combinations probably exist throughout the world.

### 5a. Has Cumbria been associated with plate movements in the past?

This is perhaps the most exciting piece of geological supposition we have ever considered. Much of the existing evidence [7] in the Lake District rocks could support the theory that we are living at a point on the Earth's surface where **two plates collided** head-on. If this is true, then the new exciting theories have a great relevance to us.

### 5b. What evidence is there to support this theory?

The distribution of land and sea on the Earth's surface has been changing slowly but constantly throughout geological time. At the start of the Ordovician Period about 530 million years ago much of the area presently covered by the British Isles was covered by a deep ocean trough or geosyncline. Iapetus is the name we now give to this ocean. Recent research indicates that the Iapetus Ocean floor was spreading outwards from a central ridge, at a time when the North and South Continents were moving towards each other.

Fossil content in rocks from Southern Scotland, the Lake District and North Wales suggest that the two shorelines ran approximately parallel to each other. Certain fossils species — e.g. graptolites — are common to rocks of the same age in all three localities. Other fossils, however, are restricted to only one locality, i.e. either north or south of the Lake District. Research on the habits of these fossils suggests that the graptolites floated on the surface of the sea and

(5) The Atlantic Ocean floor is spreading and pushing America apart from Africa. Similar movements are happening in other parts of the world. The recent earthquake disasters in Northern Italy and USSR are probably associated with the collision of plates. The "Ring of Fire" which surrounds the Pacific Ocean is due to the ocean floor descending into the molten interior.

(6) You get the same effect by pushing a few sheets of paper together. Your hands represent the plates and the paper represents the sediment.

(7) Refs. Fitton & Hughes (1970); Moseley (1976); Phillips et al (1976); Wright (1976); Sugisaki (1976).

| (i). | (ii). |
|---|---|
| 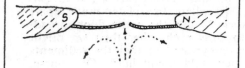  ORDOVICIAN GEOSYNCLINE BASED ON FOSSIL EVIDENCE, ETC. | 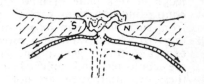  NORTH AND SOUTH COASTS MOVE TOWARDS EACH OTHER. GEOSYNCLINE CLOSES UP IN SILURIAN TIMES. |
| (iii). | (iv). |
| 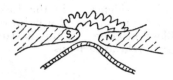  SUBDUCTION ENDS AND THE TWO COAST-LINES MEET AS THE CONTINENTAL MASSES COLLIDE. |   LOCALISED HEAT HELPED BY FRICTION - IN THE BENIOFF ZONE OF THE SUBDUCTING OCEAN FLOOR ROCKS. |
| (v). | (vi). |
|   GRANITE PLUTON IS LESS DENSE THAN THE SURROUNDING ROCKS, THEREFORE IT STARTS TO RISE TO THE SURFACE. CRUSTAL ROCKS ABOVE ARE ARCHED, WEAKENED, FOLDED AND FAULTED BY RISING PLUTON(S). |   PRESENT DAY POSITION IS REACHED WHERE PLUTON RISES AND THE ORIGINAL COVER-ROCK HAS BEEN REMOVED BY EROSION. (IT IS LIKELY THAT THE GRANITE WAS FORMED AS A SERIES OF PULSES) * |
| THIS DOES NOT CONTRADICT ANY OF THE OLDER IDEAS OF THE FORMATION OF THE LAKE DISTRICT ~ IT DOES RENDER SOME OBSOLETE IN ATTEMPTING A MORE FUNDAMENTAL EXPLANATION. | HYPOTHESISED ORIGIN OF THE LAKE DISTRICT GRANITES |

**Fig. 5**

therefore have a comparatively widespread distribution. Trilobites and brachiopods tend to be shallow water dwellers and were less able to migrate across the deep water trough; instead they moved **along** the coast. This explains why there is close similarity between species from Scotland, North America and Northern Europe, whilst they do not relate as closely to those of the same age from North Wales. As the ocean floor spread towards the coast of the "old Europe", it was forced downwards back into the Earth's interior. This is called a subduction zone.

Earlier ideas in the 1950s about the origins of the Lake District centred around the ocean floor. From at least as early as Cambrian times the Iapetus Geosyncline was gradually being filled with fine sediments which were being washed in from the Continental masses. It was suggested that the weight of these sediments caused the ocean floor to sag and draw the two coast-lines together. The ocean floor spreading/plate tectonic theory presents an alternative explanation for the movement of these Continental masses.

Thus we can establish that the Iapetus Geosyncline was slowly closing up and an estimated rate of 3cm each year has been suggested. By the end of the Ordovician times the fossil faunas, which had been restricted to opposite shore-lines, were able to mix. During mid-Ordovician times when subduction of the ocean-floor under the continental rocks was at its peak, the Borrowdale Volcanic Group of rocks were extruded onto the surface. These rocks consist of over 5000 metres[8] of fine grained igneous rocks which were poured out from a line of shallow-water volcanoes which probably ran along the coast. By late Ordovician times the North and South Continents were very close together and subduction had finished. It was now that the South Lakeland sedimentary limestone, shales and greywackes were formed. By the end of Silurian times the two approaching continents had collided and were fused together during the Caledonian orogeny. The Ordovician and Silurian rocks were greatly contorted and altered, resulting in the type of structures exhibited in the new road cuttings in the Tebay Gorge and Lune Valley. Whilst this was going on at the surface, the great granite batholith (pluton) was being emplaced in the lower parts of the continental margin.

## Explanation of Figure 4.

(i) Ocean floor moving southwards is forced under plate S. As the leading oceanic crust descends into the asthenosphere it is melted, only to be re-cycled elsewhere by the convection currents in the lava. It is these convection currents which are largely responsible for the plate movements in the first instance. This process is called subduction.

(ii) The descending rock is affected by (a) general heat due to the transfer of heat from the Asthenosphere, and (b) localised heat due to friction where it comes into contact with the underneath of plate S.

(8) Some measurements are given in metric and others in "the old way" so as to please both types of readers.

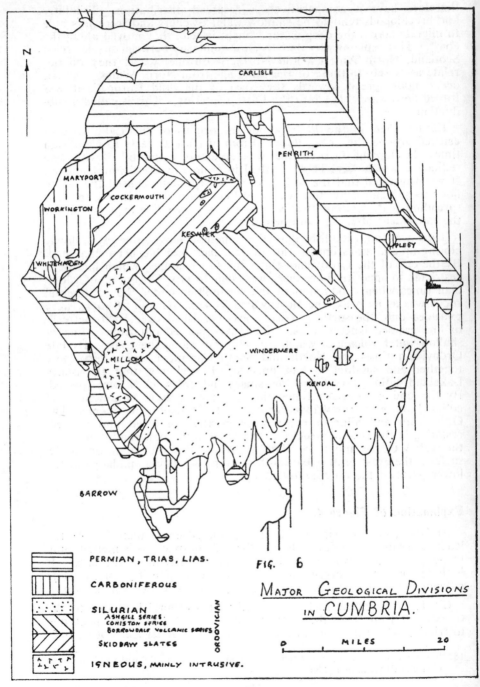

N

CARLISLE

PENRITH

MARYPORT

COCKERMOUTH

WORKINGTON

KESWICK

APPLEBY

WHITEHAVEN

WINDERMERE

MILLOM

KENDAL

BARROW

PERMIAN, TRIAS, LIAS.

CARBONIFEROUS

SILURIAN
ASHGILL SERIES
CONISTON SERIES
BORROWDALE VOLCANIC SERIES

SKIDDAW SLATES

ORDOVICIAN

IGNEOUS, MAINLY INTRUSIVE.

FIG. 6

MAJOR GEOLOGICAL DIVISIONS
IN CUMBRIA.

0          MILES          20

14

(iii) The effect of the intense local heat is to cause hotspots which form pockets of molten rock in areas of less mobile material.

(iv) These pockets of molten rock may be formed by rock from (a) the descending oceanic crust and, (b) the neighbouring minerals in the lower lithosphere and asthenosphere.

(v) The pockets of molten rock form the **Granite Pluton** which underlies most of Cumbria[9].

(vi) Such a situation constitutes a **weak point** in the crust of the Earth and stresses create cracks and fissures through which molten rocks are injected, often to be extruded onto the surface of the Earth as volcanic eruptions. The diagram shows how the volcanoes might be placed into three types. As each type derives its lava from a different source then the chemical composition of its lava could be different. These three distinct lava types are :

  (a) **Tholeiitic basalt** (augite-andesitic lava).
  (b) **Calc-alkaline lava** (calcium rich ferro-magnesium minerals, e.g. hornblende, augite, feldspar).
  (c) **Alkaline lava** (sodium and potassium feldspars, mica, amphiboles, pyroxenes and feldspar).

This is the sort of information which we look for when analysing the rocks from extinct volcanoes, and as a result we are able to construct diagram 5.

(vii) The effect of gravity on the inside of the Earth is such that it attempts to re-arrange the minerals with the heaviest ones at the centre (logical when you think that gravity is responsible for the weight of things!). The pluton of **granite** is, however, **less dense** than the surrounding rocks, and therefore it tries to rise upwards. This is similar in principle to holding an inflated balloon at the bottom of a bath of water. The balloon will want to rise and as soon as you let go, it will rapidly rise to the surface of the water. Similarly with the granite pluton; it is trying to rise slowly, though not as fast as the balloon in the bath!

(viii) An effect of this buoyant granite mass is to arch up the rocks above it. This movement causes the outer surface of the crustal rocks to be stretched and therefore be more readily weathered by the elements.

(ix) The present outcrop patterns are best seen on the 1 inch: 4 miles Geological Map Sheet 3 (1959), and the available 1 inch : 1 mile Geological Survey Maps. Almost all books on the geology of the Lake District comment at length on the dome structure flanked by successive strata of fractured Carboniferous younger rocks. This arrangement could easily occur in direct consequence to the rise of a large granite pluton.

(9) It is important to realise the time scale in these happenings—they don't happen overnight, but take millions of years.

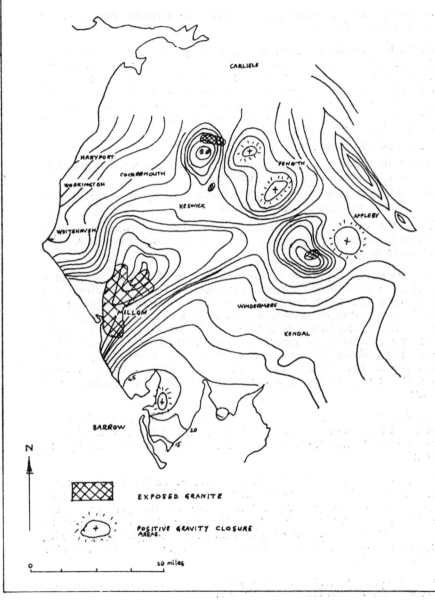

Fig 7  Gravity measurements can be plotted on a map as 'force-lines'

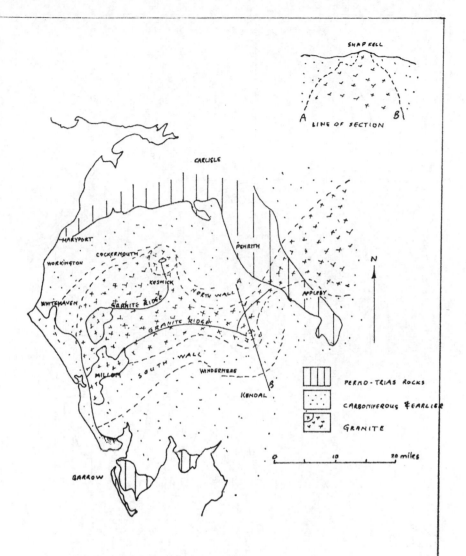

## Fig. 8. The Size of Hidden Granite Batholith

Using the gravity measurements and the relative positions of the known granite outcrops, Bott constructed a map to show what he considers to be the probable extent of the granitic mass which underlies much of Cumbria. The edges of the mass slope outwards.

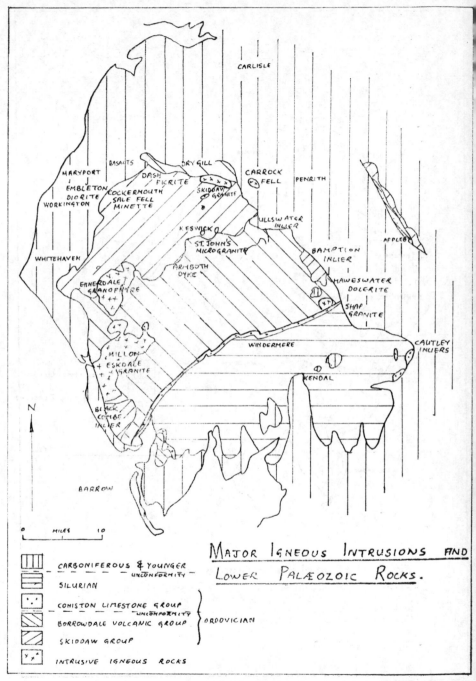

CARLISLE

MARYPORT
BASALTS
DRY GILL
CARROCK FELL
PENRITH

EMBLETON DIORITE
DASH FICRITE
SKIDDAW GRANITE

WORKINGTON
COCKERMOUTH SALE FELL MINETTE

KESWICK

WHITEHAVEN

ST. JOHN'S MICROGRANITE
ULLSWATER INLIER

BAMPTION INLIER

ARMBOTH DYKE

ENNERDALE GRANOPHYRE

HAWESWATER DOLERITE

SHAP GRANITE

WINDERMERE

APPLEBY

CAUTLEY INLIERS

MILLOM
ESKDALE GRANITE

KENDAL

BLACK COMBE INLIER

N

BARROW

0    MILES    10

MAJOR IGNEOUS INTRUSIONS AND
LOWER PALÆOZOIC ROCKS.

| | CARBONIFEROUS & YOUNGER |
| --- | --- |
| | — UNCONFORMITY — |
| | SILURIAN |
| | CONISTON LIMESTONE GROUP |
| | — UNCONFORMITY — |
| | BORROWDALE VOLCANIC GROUP | ORDOVICIAN |
| | SKIDDAW GROUP |
| | INTRUSIVE IGNEOUS ROCKS |

**Fig. 9**

## 6.  What of the future?

A good question!  There is no reason why the dynamic processes, which are an ongoing part of the geological story of the Earth, should cease because we have recently discovered knowledge of them. Indeed, if the current idea of the spread of Gondwanaland — the hypothetical southern hemisphere "super-continent" — is true, then it could be merely a matter of time before all the pieces meet round the other side of the Earth! Likewise, if the granite pluton under Cumbria has started to rise, then surely it will continue to rise until it has achieved a state of equilibrium (i.e. balance) by floating on the denser surrounding rocks. Recent gravity measurements [10] over the north of England have shown that the Lake District granites are connected to the North Pennine granites by a hidden, high-level granite ridge which runs W - E from Shap under the Vale of Eden towards Knock Fell.

As the granite pluton continues to rise it will place increasing stress on the surface rocks. This will cause folding [11] and faulting and earthquakes may occur where there is a comparatively sudden release of pressure.  Whether we might expect a renewal of volcanic activity associated with uplift is debatable.  As the pluton becomes more distant from its sources, and moves away into the upper levels of the lithosphere, then it has lost much of its initial heat.  In any event, hot springs would be a timely indication of a renewal of vulcanism in the district.

## 7.  Where are these igneous "clues" for the granite pluton/plate collision theory?

The rocks collectively called the **Borrowdale Volcanic Group** (or Series) provide much of the evidence.  There are two contrasting outcrops :

(i) The **Southern Outcrop** is the larger, and forms the mountains of Central Lakeland (Scafell).  The main rock types are all igneous — tuffs, ignimbrites, andesite lavas, ashes and rhyolite flows.  These igneous rock terms are only generalisations, and a more exact definition is too technical to be included here.

(ii) The **Northern Outcrop** stretches in a narrow band from the River Caldew in the east to the village of Bothel in the west.  The small inliers of Eycott Hill and Greystoke Park form part of this outcrop.

The rocks in the two outcrops, whilst both being **igneous** in origin and of similar age, are of different types : The **Southern Outcrop rocks** are garnet bearing and fine grained, whilst the **Northern Outcrop** consists mainly of basalts and other low-quartz bearing mafic rocks which

(10) **Ref. Bott, M.H.P. (1974).**

(11) **Visit** the M6 motorway cuttings in and near the Tebay Gorge to see good examples of folding and faulting.

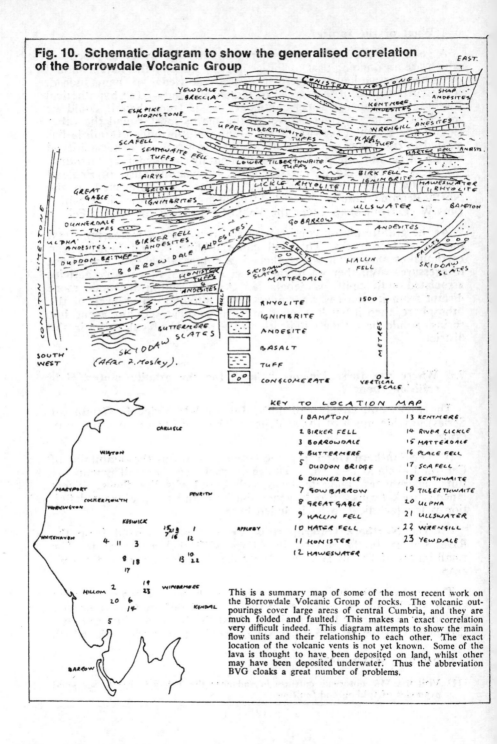

# Fig. 10. Schematic diagram to show the generalised correlation of the Borrowdale Volcanic Group

EAST.

CONISTON LIMESTONE
YEWDALE BRECCIA
SHAP ANDESITES
ESK PIKE HORNSTONE
KENTMERE ANDESITES
UPPER TILBERTHWAITE TUFFS
WRENGILL ANESITES
SCAFELL
SEATHWAITE FELL TUFFS
PLACE FELL TUFF
LOWER TILBERTHWAITE TUFFS
HARTER FELL ANDSTS.
AIRYS BRIDGE IGNIMBRITES
BIRK FELL IGNIMBRITE
LICKLE RHYOLITE
GREAT GABLE
HAWESWATER RHYOLITE
DUNNERDALE TUFFS
ULLSWATER
BAMPTON
ULPHA ANDESITES
BIRKER FELL ANDESITES
GO BARROW
ANDESITES
DUDDON BR: TUFF
BORROW DALE ANDESITES
FAULTS
DUDDON BR: TUFF
HONISTER TUFFS & ANDESITES
FAULTS
SKIDAW SLATES
MATTERDALE
HALLIN FELL
SKIDOAW SLATES
BUTTERMERE SLATES
SKIDOAW

SOUTH WEST

(After J. Mosley).

|  |  |
|---|---|
| ▦ | RHYOLITE |
| ∿ | IGNIMBRITE |
| ∴ | ANDESITE |
| ▤ | BASALT |
| ∴ | TUFF |
| ○○ | CONGLOMERATE |

1500

METRES

0

VERTICAL SCALE

CONISTON LIMESTONE

## KEY TO LOCATION MAP

| | |
|---|---|
| 1 BAMPTON | 13 KENTMERE |
| 2 BIRKER FELL | 14 RIVER LICKLE |
| 3 BORROWDALE | 15 MATTERDALE |
| 4 BUTTERMERE | 16 PLACE FELL |
| 5 DUDDON BRIDGE | 17 SCAFELL |
| 6 DUNNER DALE | 18 SEATHWAITE |
| 7 SOW BARROW | 19 TILBERTHWAITE |
| 8 GREAT GABLE | 20 ULPHA |
| 9 HALLIN FELL | 21 ULLSWATER |
| 10 HATER FELL | 22 WRENGILL |
| 11 HONISTER | 23 YEWDALE |
| 12 HAWESWATER | |

CARLISLE
WIGTON
MARYPORT
COCKERMOUTH
WORKINGTON
PENRITH
KESWICK
15 21 3
7 16  1
APPLEBY
12
WHITEHAVEN
4  11  3
8  18
10
13  22
17
19
MILLOM  2  23  WINDERMERE
20  6
14  KENDAL
5
BARROW

This is a summary map of some of the most recent work on the Borrowdale Volcanic Group of rocks. The volcanic out-pourings cover large areas of central Cumbria, and they are much folded and faulted. This makes an exact correlation very difficult indeed. This diagram attempts to show the main flow units and their relationship to each other. The exact location of the volcanic vents is not yet known. Some of the lava is thought to have been deposited on land, whilst other may have been deposited underwater. Thus the abbreviation BVG cloaks a great number of problems.

often contain big feldspar phenocrysts (Eycott Hill lava). Even with the naked eye, it is not too difficult to distinguish between the porphyritic, dark-coloured basaltic andesites of the **North,** and the pale, seldom porphyritic andesites of the **South.**

In conclusion, it might be supposed that the two main types of the Borrowdale Volcanic Group rock suites were erupted at the junction of two colliding plates [12] early in the Caledonian earth-moving period in late Silurian times — about 440 million years ago — the Eycott Lavas being marginally older than those further south.

8. **Are there any other igneous rocks in Cumbria other than those associated with the Borrowdale Volcanic Group?**

Yes; the BVG are called **Extrusive** rocks because they were forced out onto the Earth's surface where they cooled rapidly into fine-grained suites of rocks. There are two other groups of igneous activity represented in local rocks.

(i) **The Intrusives :** These are associated with molten material which cooled much more slowly, deeper inside the Earth's crust. The granitic plutons referred to earlier come under this heading. The petrology of these intrusives is extremely complex as one series has a large degree of overlap with another, and their relative ages are still not yet fully understood. The main types are listed in the key of the 1 inch: 4 miles Geol. Map. They include Gabbro, Dolerite, Diorite, Quartz-felsite, Granophyre, and Granite. Some of these are massive bodies of rock, whilst others occur in dykes. These are infilled cracks which radiate out like the spokes of a wheel, usually from an intrusive mass. The cracks are filled with solidified lava. Lamprophyre is one of the more common rocks of this type. Intrusive rocks are always newer than the rocks into which they were intruded.

(ii) **The Contemporaneous :** Igneous rocks of this type in Cumbria are two-fold. One is BVG which we have already mentioned. The other comprises the **Basaltic Lavas** of the **Carboniferous** Period. Contemporaneous means that these lavas are found in sequence with the other rocks which were being formed at that time. Igneous rocks are usually placed together at the foot of the geological sequence on the key of IGS survey maps.

9. **What is the significance of this division of types of igneous rocks?**

As mentioned earlier, it is by careful recording and analysis of our igneous rocks that we can postulate a theory for the origin of much of the present landscape. Many factors control the type of igneous rock formed, but perhaps the key one is the **rate of cooling** of the molten material. The same minerals occurring in similar proportions

(12) Such periods of vast earth movements are called orogenies.

21

# TYPICAL THIN SECTION VIEWS of SIX DIFFERENT IGNEOUS ROCKS FROM CUMBRIA.

## 1. GABBRO

AUGITE AND LABRADORITE FELDSPAR. SMALL AMOUNTS OF ALTERED MICA AND QUARTZ. TYPICAL OF CARROCK FELL AREA. FRACTURES IN AUGITE GIVE 'HERRING BONE' APPEARANCE

## 4. BASALT

NUMEROUS SMALL LATHS OF FELDSPAR, OFTEN SURROUNDING SMALL BUNCHES OF LARGER FELDSPARS. GROUNDMASS ALSO CONTAINS GRAINS OF OLIVINE AND AUGITE.

## 2. SHAP GRANITE

LARGE CRYSTALS OF FELDSPAR IN GROUND-MASS OF FINER GRAINED QUARTZ, MICA AND FELDSPAR.

## 5. RHYOLITE

FINE GRAINED GROUNDMASS, OFTEN CONTAINING SINGLE PIECES- BITS OF QUARTZ FOR EXAMPLE. SHOWS SIGNS OF BANDING. MAINLY VERY FINE QUARTZ AND FELDSPAR.

## 3. DOLERITE

LATHS OF FELDSPAR. OPHITIC TEXTURE GROUNDMASS OF FINE AUGITE AND HORNBLENDE

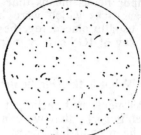

## 6. TUFF

EXTREMELY FINE GRAINED AND UNIFORM TEXTURE. SIMILAR IN SOME WAYS TO RHYOLITE. THIS SECTION IS TYPICAL OF THE LANGDALE STONE AXE FACTORY.

**Fig. 11**

can result in vastly different types of rock depending on where and how they cooled. To generalise : **Extrusive** volcanic rocks cool very rapidly and have a glassy appearance;[13] **Intrusive** rocks cool less quickly and have a mixture of small, medium and large crystals (microcrystalline and porphyritic); and **Plutonic** rocks cool very slowly and have a totally crystalline texture. Diagram 12 shows how different igneous rocks can have the same mineral composition.

### Explanation of Diagram 12 :

**Reading vertically** [14]

**Column 1** gives the three main types of igneous rocks and their texture as seen in hand specimens.

**Column 2:** These are **Felsic** rocks and used to be called acid rocks. They are the pale rocks and have the greatest amount of quartz. Notice that this table explains how rhyolite from Langdale, micro-granite from Carrock Fell, and granite from Shap, all have similar mineral composition. The fact that hand specimens of each bear no resemblance to each other is because they cooled at different rates.

**Columns 3, 4 and 5** are used in the same way as Column 1. Notice how the percentages and composition of constituent minerals changes with each column. The hand specimens get darker as the fifth column is approached.

### Reading horizontally

**Column 6** shows that fine-grained igneous rocks need not have the same mineral content.

**Columns 7 and 8** show the same as column 6 for Intrusive and Plutonic rocks respectively.

**Column 9** is the percentage by volume of minerals which are distributed throughout the rock types above.

As stated earlier, the petrology of the Lake District Igneous rocks is rather complex. If you want to make a collection of different rock types, consult the Solid and Drift editions of the 1″ : 1 mile geological map sheet 23 — Cockermouth — and you will be off to a good start [15]

---

(13) Industrial slag has been tipped in various parts of Cumbria. It very closely resembles obsidian. There are only minute traces of volcanic glass reported from the Lakeland igneous rocks.

(14) This is a very generalised diagram. Igneous petrology text-books contain more detailed and accurate graphs and charts.

(15) Some of the 1″ : 1 mile Geological maps of the Lake District are out of print at the moment. The 1″ : 4 miles has just been reprinted by H.M.S.O.

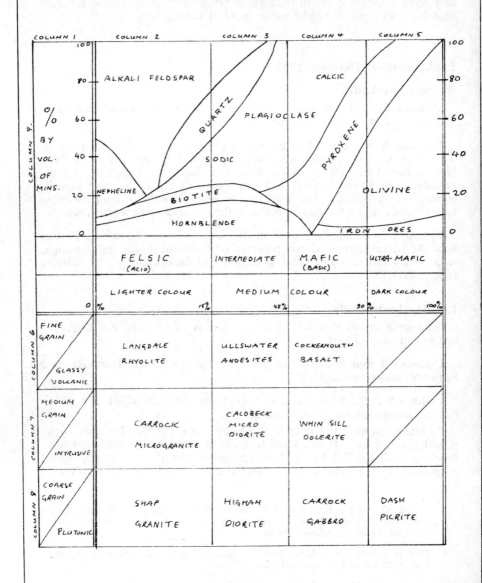

# CLASSIFICATION OF THE MAIN IGNEOUS ROCKS IN LAKELAND.

**Fig. 12**

## 10. How shattered and strained are the Cumbria Rocks?

The crust of the Earth is under a constant pressure [16] and evidence suggests that there have been several major periods of earth movement which have affected Cumbria. Fig. 13 shows the general trends of folds and faults in the British Isles. Notice how the most affected areas tend to be associated with the harder rock outcrops. Much recent off-shore work [17] on fault detection by elaborate gravity surveys has helped in the accurate mapping of the sea-bed. Much of this exploration was prompted by the prospect of oil and gas finds.

A Fault is the general name given to a point in the Earth's crust where the rock has fractured under immense stress and strain [18]. There are a variety of faults in Cumbria: normal, reverse, thrust, wrench. Each describes the particular movement of one side relative

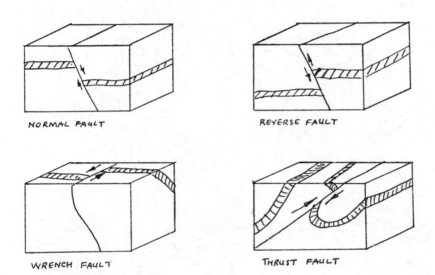

NORMAL FAULT     REVERSE FAULT

WRENCH FAULT     THRUST FAULT

to the other. Figs. 14, 16 and 17, show the locations of the major faults in Cumbria, belonging to different periods of geological time and hence being of different ages. Obviously a fault is always more recent than the rocks in which it occurs.

(16) The pressures within the Earth's crust affect the formation of volcanoes. Some like Krakatoa erupt with tremendous force, estimated at 160,000 Hydrogen bombs exploding simultaneously. Others which extrude onto the sea-bed can cause the ocean to boil, whilst others can cause islands to be formed (Sturtsey, near Iceland) overnight.

(17) The Irish sea-bed between the Lake District, the Isle of Man and Ireland has now been geoligically mapped.

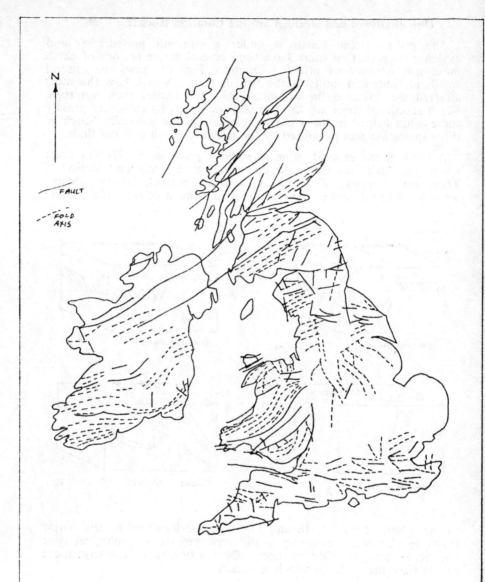

N

FAULT

FOLD
AXIS

Fig 13:    The Main Folds and Faults in the
British Isles.

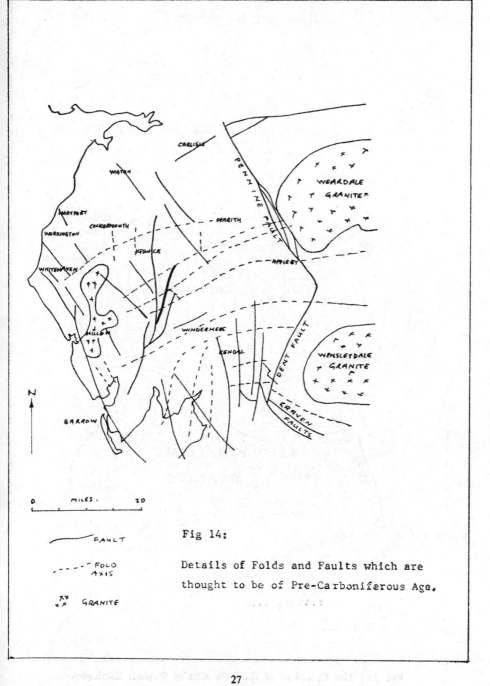

Fig 14:

Details of Folds and Faults which are
thought to be of Pre-Carboniferous Age.

27

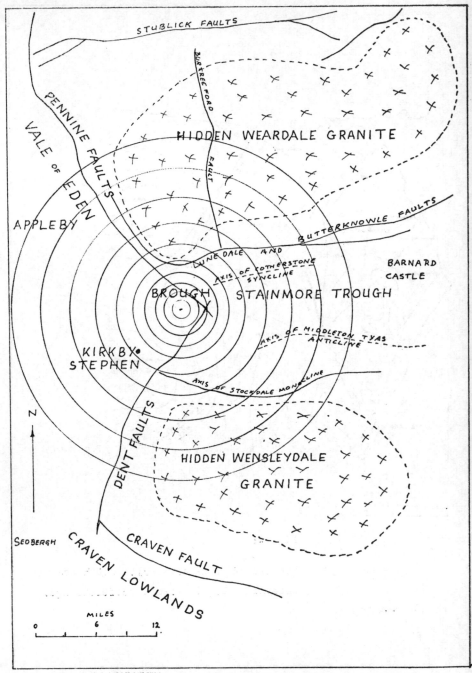

**Fig. 15: The Epicentre of the 1970 Kirkby Stephen Earthquake.**

**The Kirkby Stephen Earthquake :** At 9-10 p.m. on Sunday, 9th August, 1970, local people — especially those living in the eastern part of Cumbria — witnessed what transpired to be one of the strongest **earth tremors** ever recorded in this country. Two contrasting explanations have been given : that a new, weak plane had developed trending from the North-East towards Morecambe Bay, or that there had been renewed activity along an existing fault-line. Most now agree that the second alternative is more likely, and that the **epicentre** of the tremor was at a point about 15 kilometres below the Brough-Kirkby Stephen-Sedbergh area. This would mean very near the junction of the Pennine and Lunedale Faults, and along the Dent Fault. An intensity of V on the modified Mercalli Scale was recorded. This equates with 4.9 on the Richter Earthquake Scale, which makes it the strongest tremor on land in Britain since the scale was adopted in 1935. There is little doubt that the tremor resulted from an adjustment to a build-up of stress within the Earth's crust. Although this area is undoubtedly affected by the current Alpine Orogeny it is more likely that the Kirkby Stephen tremor was attributable to stress caused by the uplift of the granite ridge which is thought to exist under the area, running between the Lakeland Granite (Devonian Age) in the west and the North Pennine Granites (Weardale — Devonian Age; Wensleydale —? Pre-Cambrian Age) to the east. Such local stress fields will, no doubt, be further aggravated by the subsiding North Sea Basin.

**Folds** often result when the rate of pressure in the Earth's crust is slow enough to allow rocks to readjust without breaking. The two basic fold types are anticlines (upfolds) and synclines (downfolds). Perhaps the best place to observe the effects of folding is where sedimentary rocks have been affected. Usually such rocks were originally deposited as **horizontal** strata on the sea floor. In Cumbria they are often found steeply inclined, having been uplifted and tilted by Earth movements in the past (and present!). Fig. 14 shows that there is a dominant trend of folds having a NE/SW axis. In section these would resemble a series of ripple marks. One of the finest local exposures of folding and faulting can be found in the Tebay Gorge where recent motorway construction work has revealed a whole range of structures in the Ordovician/Silurian sedimentary rocks.

### 11. Is there an equally wide variety of Sedimentary rocks to be found in Cumbria?

It all depends upon what you base your classification of rocks. Two methods come immediately to mind — Stratigraphical and mode of formation.

(i) **Classification by Stratigraphical Succession.** This simply means looking at the rock types in the order that they are found in the

(18) Many faults in Cumbria have been mineralised.

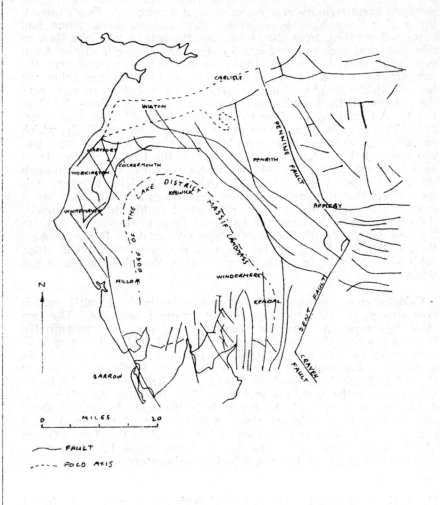

CARLISLE

WIGTON

MARYPORT

COCKERMOUTH

WORKINGTON

WHITEHAVEN

PENNINE FAULT

PENRITH

APPLEBY

EDGE OF THE LAKE DISTRICT

KESWICK

MASSIF LANDMASS

WINDERMERE

KENDAL

DENT FAULT

N

MILLOM

BARROW

CRAVEN FAULT

0        MILES        20

——— FAULT
----- FOLD AXIS

Details of faults which are thought
to be of Carboniferous and later age.

**Fig. 16**

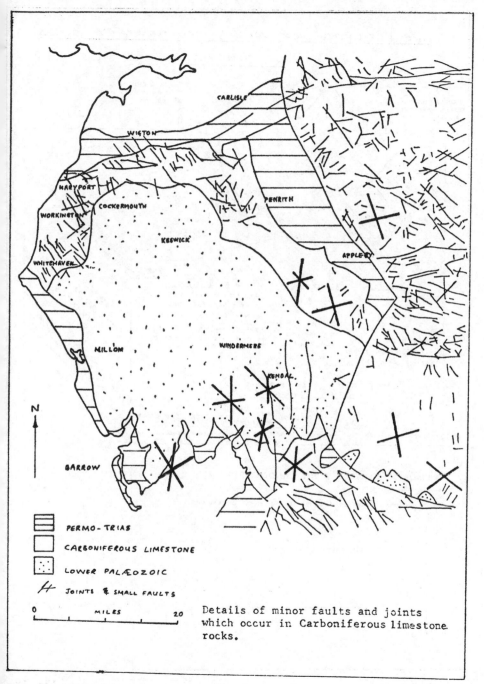

CARLISLE

WIGTON

MARYPORT

COCKERMOUTH

WORKINGTON

KESWICK

PENRITH

APPLEBY

WHITEHAVEN

MILLOM

WINDERMERE

KENDAL

N

BARROW

| | PERMO-TRIAS |
| | CARBONIFEROUS LIMESTONE |
| | LOWER PALÆOZOIC |
| | JOINTS & SMALL FAULTS |

0   MILES   20

Details of minor faults and joints
which occur in Carboniferous limestone
rocks.

**Fig. 17**

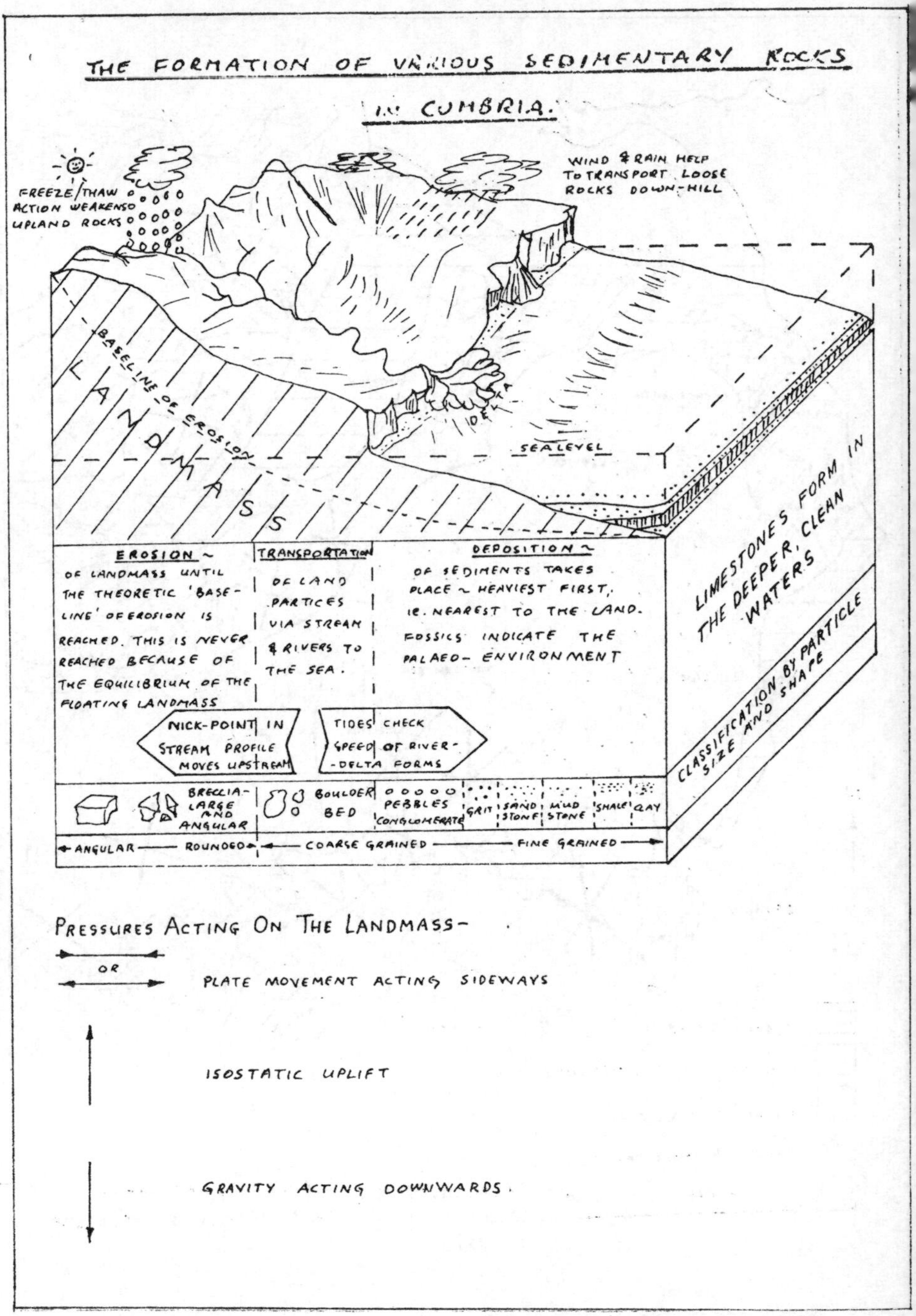

# THE FORMATION OF VARIOUS SEDIMENTARY ROCKS

## IN CUMBRIA.

FREEZE/THAW ACTION WEAKENS UPLAND ROCKS

WIND & RAIN HELP TO TRANSPORT LOOSE ROCKS DOWN-HILL

BASELINE OF EROSION

LANDMASS

DELTA

SEALEVEL

LIMESTONES FORM IN THE DEEPER, CLEAN WATERS

CLASSIFICATION BY PARTICLE SIZE AND SHAPE

**EROSION ~**
OF LANDMASS UNTIL THE THEORETIC 'BASE-LINE' OF EROSION IS REACHED. THIS IS NEVER REACHED BECAUSE OF THE EQUILIBRIUM OF THE FLOATING LANDMASS

**TRANSPORTATION**
OF LAND PARTICES VIA STREAM & RIVERS TO THE SEA.

**DEPOSITION ~**
OF SEDIMENTS TAKES PLACE ~ HEAVIEST FIRST, ie. NEAREST TO THE LAND.
FOSSILS INDICATE THE PALAEO-ENVIRONMENT

NICK-POINT IN STREAM PROFILE MOVES UPSTREAM

TIDES CHECK SPEED OF RIVER-DELTA FORMS

| | | | | | | | | |
|---|---|---|---|---|---|---|---|---|
| BRECCIA-LARGE AND ANGULAR | BOULDER BED | PEBBLES CONGLOMERATE | GRIT | SAND STONE | MUD STONE | SHALE | CLAY |

← ANGULAR — ROUNDED → ← COARSE GRAINED — FINE GRAINED →

## PRESSURES ACTING ON THE LANDMASS-

← → OR ← → PLATE MOVEMENT ACTING SIDEWAYS

↑ ISOSTATIC UPLIFT

↓ GRAVITY ACTING DOWNWARDS.

**Fig. 18**

ground (i.e. related to the order in which they were deposited). The problem here is that there is bound to be some duplication as similar conditions have existed in our area on more than one occasion in the past. Fig. 19 shows the main sedimentary rock types which were formed at different periods of geological time. Even in such general terms, sandstone occurs four times, shales four and limestone twice. Therefore, for the non-specialist geologist, it is suggested that a simple classification based on the relative position in the Stratigraphical Column is not too helpful.

(ii) **Classification by Mode of Formation.** This is probably more helpful. Consider two basic modes of origin — mechanical and organic. **Mechanically formed sedimentary rocks** are formed as a result of weather action wearing down the land. The main agents of this erosion, transportation and deposition process are gravity, rain, snow, wind, sun, and running water. A simple typology for mechanically formed sedimentary rocks is by using grain size of the constituent particles. Fig. 18 shows a theoretical case of erosion. Transportation and deposition are kept distinct, but in fact these three can occur at any point on the continuum wherever the necessary conditions are present. The distinction between the rock types is rather blurred — i.e. coarse grit, grit, fine grit, coarse sandstone, sandstone — and the average diameter of the grains is often used as the criterion [19].

$$\text{Conglomerates } 2mm + \qquad\qquad \text{Silts } \frac{1}{16} - \frac{1}{256} \text{ mm}$$

$$\text{Sandstones } 2 - \frac{1}{8} \text{ mm} \qquad\qquad \text{Clays less than } \frac{1}{256} \text{ mm}$$

When the grains are assorted in size and shape they form a rock called Greywacke — a common rock type in geosynclines. Many of our Silurian rocks in South Cumbria are varieties of this type.

By comparing diagram 18, showing the **conditions** of formation, with diagram 19 showing the **age** of formation, we can deduce the nature of Cumbria in the past and trace successive developments. Once you get the feel of this you are on the way to becoming a "geologist". Soon you will be able to "read" rocks just as a musician reads music. It is perhaps this sort of observation and deduction which gives an amateur geologist an insight into the skills, satisfaction and frustrations of the professional.

The Institute of Geological Sciences (IGS) Geological Survey Maps represent the present published geological knowledge of our area. Some maps are accompanied by a detailed sheet memoir, e.g. 1″ Cockermouth and 2½″ Cross Fell sheets.

---

(19) You need pretty good eyes for this! Some quarrymen distinguish between siltstone and mudstone by grinding a small sample between their teeth. This is not the usually accepted use of the "taste test".

## Figure 19

| Geological Period | Age in millions of years | Sedimentary rock type | Exposure Location |
|---|---|---|---|
| Recent | | Blown sand | Ravenglass |
| Recent | | Raised Beach deposits | Silloth |
| Recent | 2 | River Gravels & Alluvium | Bassenthwaite Lake |
| Jurassic | 190 | Lower Lias Clays | W. Carlisle (borehole only) |
| Triassic | | Keuper Marl & Stanwix shales | Carlisle |
| Triassic | 225 | St. Bees Sandstone & shales | St. Bees Head |
| Permian | | Penrith Sandstone & Brockram | Appleby |
| Permian | 280 | Magnesium Limestone | Salton Bay, Whitehaven |
| Upper Carbon | | Coal Measure sandstone, shales & clays | Whitehaven/Workington |
| Middle Carbon | | Millstone Grit | Alston |
| Lower Carbon | | Main Limestone Groups | Great Asby |
| Lower Carbon | 345 | Basalt Conglomerate & Sandstones | Shap Wells |
| Silurian | 440 | Bannersdale 'Slates'; Coniston Flags; Stockdale Shales | Windermere |
| Ordovician | 530 | Coniston Limestone | Coniston |

34

## 12. If there are igneous and sedimentary rocks present in Cumbria, has one caused the metamorphism of the other?

Any rock which has been subjected to excessive heat/pressure/ solution can be metamorphosed. By their very nature, metamorphic rocks often contain an impressive range of unusual minerals, but don't be disappointed at not finding them all. Such minerals are only detectable by sophisticated electronic devices and are often present in micro-concentrations. For the sake of simplicity only two types of metamorphic rock will be mentioned here — one resulting from the metamorphism of an igneous rock, the other from a sedimentary rock.

(i) **Shap Blue Quarry Hornfels** is a metamorphic rock which was formed by the heat of the Shap Granite altering the neighbouring andesites and rhyolites of the BVG. The result is a hard dark grey/ blue fine-grained rock which contains concentration of certain minerals — garnet, iron pyrites, epidote and calcite. These have resulted from the reorganisation of the original constituents into new mineral assemblies.

(ii) **Sinen Gill Slates** were formed when the Skiddaw Slates were intruded by the Sinen Gill Granite, being part of the Skiddaw Granite suite. Obviously, the greatest heat was at the point of contact between the granite and the country-rock, and the intensity of heat decreased with distance from the point of intrusion. Therefore we would expect to find the Skiddaw Slates reacting differently according to the degree of heat they received. A generalisation is shown in diagram 20. A walk from the Lake District National Park Centre at the old Threlkeld Sanatorium to Sinen Gill goes across text-book exposures of regional metamorphism grading into contact metamorphism [20].

The inter-bedded basaltic Carboniferous lavas of the Cockermouth area also had an effect on their country rocks, though these lavas were almost cold when compared with the intrusive bodies which underlie Ennerdale, Eskdale, Skiddaw, St. John's, Shap and Carrock.


## 13. What is the age of the Cumbria granite?

It is postulated that all the Lake District granites connect underground. The most conclusive proof of age is found in the Basal Carboniferous Pebble Bed which is exposed between Shap Wells and Tebay. The Shap Granite is intruded into BVG and Silurian sediments; therefore it is **newer** than Silurian age. The Pebble Bed (conglomerate) contains many pebbles of Shap Granite. Therefore the

(20) The Skiddaw Slates are not homogenous. They consist of alternating beds of fine and coarser grained sediments which range from mudstone to grit. Thus each type will react differently to the same metamorphic agents. It is deceptive therefore to regard the metamorphism of the Skiddaw Slates as a simple process with uniform results.

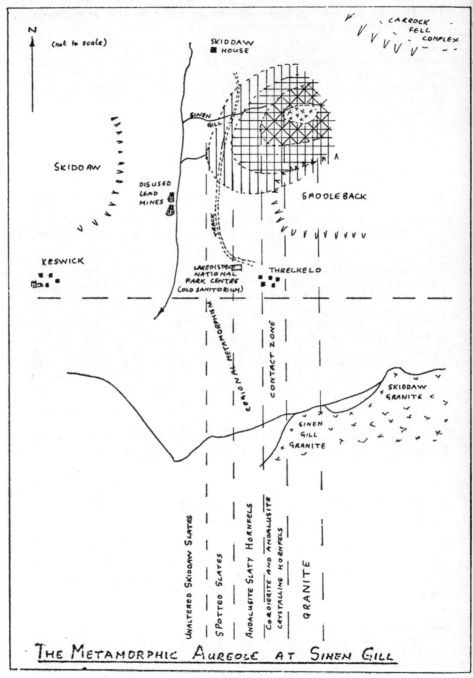

THE METAMORPHIC AUREOLE AT SINEN GILL

Fig. 20

granite must have been emplaced **before** the start of the Carboniferous period. This suggests that the Shap Granite is of **Devonian** age, a deduction more recently confirmed by radio carbon dating which puts most Cumbrian granites at about 400 million years old.

## 14. Which minerals and rocks have been economically extracted in the past?

(i) The oldest "industrial" use of stone for which we have substantial evidence is the Neolithic stone axe industry which flourished about 4000 BC. About a dozen axe factory sites have been identified in the Langdale Valley and an excellent display of maps, diagrams and specimens is available for reference at Carlisle Museum. Other Cumbrian museums also have a range of local Neolithic material on display. Evidence suggests that these "primitive" men were very highly selective in their choice of raw materials for making their "Cumbrian Clubs", and the axe factory sites appear to be restricted to a definite outcrop of fine-grained volcanic ash which suited the purpose ideally. It is thought that the Langdale factories probably had competitors in the axe-field. Petrological analyses of other axes found in Cumbria indicate that similar axes were also produced from different rock types. Another common rock used was a grey/blue, fine grained sedimentary rock. The exact source of the rock and factory sites have yet to be discovered, though the latest supposition is that it is somewhere in the South Lakeland area, possibly the Bannersdale Slates of the Howgill/Whinfell region. Work on the petrology and archaeology of stone axes is co-ordinated by the Council for British Archaeology. The British Museum (Nat. History) and a team of about six petrologists are gradually piecing together the national jigsaw.

(ii) **Metallic minerals,** or ores as they are more often called, occur in many sites within Cumbria. Such deposits can be divided into two groups based on their mode of formation :

(a) **The Vein Deposits :** These are usually metallic, sulphur-rich ores and are not too difficult to identify — e.g. iron pyrites, copper pyrites, galena, sphalerite. Where these "primary" ores have been affected by the weather, chemical alteration takes place and they take on a different appearance — e.g. gold coloured iron pyrites may turn to a rusty-looking limonite, gold coloured copper pyrites may turn to either green malachite or blue azurite; the heavy black ore of galena may turn to a heavy white crystal of cerrusite; and the zinc ore sphalerite (alias Black Jack or Zinc Blende) may tarnish from its original appearance.

Until recently such ore deposits in the Lake District and North Pennines were thought to have originated as the result of mineralised gases and hot solutions which were forced **upwards** into cracks in the Earth's crust. Often such cracks were existing faults which subsequently became mineralised. It was suspected that the source of this

## Fig. 21. The Distribution of Neolithic Age Stone Axes of possible Cumbria origin

● - GROUP Ⅵ
○ - GROUP ⅩⅤ

Such distribution maps can be misleading — rocks of a similar nature are known to exist in more than one locality. The distribution of finds tends to reflect the distribution of lookers. Several arrow-heads which have been found in Cumbria and are recorded as "flint" (from the Chalk of Cretaceous Age) are now thought to be of a more local origin, being made from local cherts.

The Council for British Archaeology has established research committees which examine the types of rock which were used in the manufacture of axes in Neolithic times. By identifying the rock type and noting the find-spot of the axe, archaeologists try to trace the trade and movement patterns of our ancestors about 6,000 years ago. Two groups of axes are suspected as originating in Cumbria. These are:—

(i) **Group 6,** being axes made from a medium grained volcanic tuff which contains small fragments of quartz and feldspar, small iron ore and epidote crystals, all set in a fine grained matrix of felsic material. The axe factories in the Scafell-Langdale area are most likely to be the source of many of these axes and several hundred "rough-outs" have been collected from screes in the Langdale Valleys. The occurrence of polished axes, however, is far less common. The largest concentrations tend to occur on the Cumbrian coast where it appears that the roughed out axes were polished with sand from the beach. A word of warning: rocks of a very similar nature to the Langdale outcrops also occur in Wales, and it is quite possible that some of the axes indicated on the map could have originated there and not in Cumbria.

(ii) **Group 15,** being axes made from a micaceous fine grained greywacke which very closely resembles the Coniston Grits of South Lakeland. The rock contains angular grains of quartz and feldspar , flakes of mica and fragments of felsite and schist, set in a cement of Chlorite, calcite and iron ore. No factory site has yet been located for these axes. Coniston was thought a likely place for manufacture, but a more recent study of cleavage of the axes and minerals tends to suggest the Whinfell Ridge/Howgill Fells as a more likely locality.

heat was a granitic mass which might conveniently underlie the mineral veins. More recent investigations of certain minerals, especially fluorspar and barytes, seem to indicate that this hot origin for the vein ores might not be quite true. We now suspect that much of Cumbria and the North Pennines are underlain by the same granite batholith which trends roughly East/West across the Vale of Eden. Whilst this granite is probably linked with the ore deposits it may well be that it was **cold** and not hot at the time of mineral deposition.

Why do we think this? The existence of Shap Granite pebbles in the conglomerates at the base of the Carboniferous Limestone rocks suggests that the granite had been emplaced, cooled down and was being weathered away at the surface before Carboniferous times. As it cuts through Ordovician and Silurian rocks, we can place this erosion in Devonian times.

A recent borehole sunk in the Appleby area has proved that a granite (Weardale Granite — overlain by the Carboniferous Limestone) also exists there at about 2,000 feet below the present surface. The existence of a thin soil layer between the granite and limestone confirms the fact that the granite in this area was exposed at the surface and was being weathered before Carboniferous times. Therefore, it is difficult now to see how in this instance the granite could have caused the hot mineralised solutions, as it had cooled even before the Carboniferous rocks had been deposited. As many of the mineral veins occur in the Carboniferous Limestones, there can be little doubt that they are at least newer than this age. It now seems likely that the minerals originated in a colder state than suspected previously.

Further evidence supports this idea for a cold origin of some local mineral veins. Specimens of vein minerals have been found to contain fluid inclusions. These are small sealed cavities containing a brine solution which became trapped as the crystals were forming. The brine solution often contains small bubbles which formed as the solution presumably cooled and contracted within the vacuum of its sealed cavity. By heating these bubbles until they disappear and noting the temperature, it is possible to discover the temperature at which the mineral crystals were originally formed. In most cases this ranges from $60°$ to $200°$ Centigrade. We already have conclusive proof that the "pure" water which circulates on and just below the Earth's surface becomes increasingly saltier (i.e. rich in sodium chloride) at depth. In some cases the brine can contain up to six times its usual concentration and be capable of holding large volumes of material in solution. Such super-rich brine solutions therefore might conceivably percolate upwards through the granites and dissolve out iron, zinc, copper, lead and other minerals which are later deposited along the sides of faults and joints in the limestones and other receptive strata above, during a series of pulses of activity.

The distribution of minerals and the orientation of the veins also suggests the existence of a rising granite mass closely linked with vein deposits. This does not contradict the cold origin of such deposits. Given the rising brine as a mechanism for dissolving, transporting and

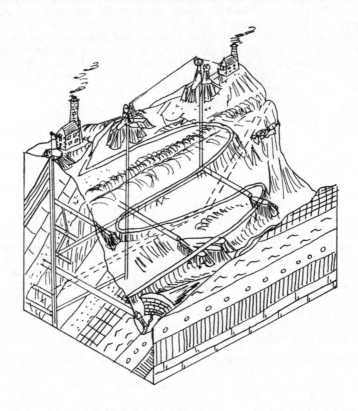

This diagram shows the most usual lay-out of one of the "old type" of metal mines, common in various parts of the country. It is typical for the extraction of vein deposits which are steeply inclined. We would expect to find primary sulphide ores in the lower sections, and secondary altered carbonate ores nearer the surface. Often, there is a zone of "secondary enrichment" somewhere near the bottom of the vein caused by the descending surface waters concentrating ores leached from above during their passage. In the olden days, when drainage problems were too difficult and expensive to overcome, many mines were abandoned before this enriched zone was reached. It is thought by some that for this reason alone, it may be financially rewarding to reopen some of the mines abandoned well over a hundred years ago.

### Explanation of Terms

**Stope** — The space left underground where the ore has been removed.

**Ore** — A rock or mineral which, when chemically treated, will yield a metal.

**Adit** — A tunnel driven from the surface to reach the ore. It is usually horizontal, and starts in a valley side. There is often one adit which is larger than the others, and is lined with stone, containing rails for transporting the ore. This is called the Haulage Adit. Somewhere under the mine workings there may be an adit which connects to a nearby stream. The idea is that the water entering the stopes will eventually drain down to the lower levels where it can be removed from the mine to the stream. This would be known as a **Drainage Adit**.

**Shaft** — A vertical hole sunk onto the orebody. Shaft used to hoist the ore from the stopes was called a Haulage Shaft; shafts used to take the men to work were called Man Shafts; shafts used to ventilate the workings were called Air Shafts.

**Cross-cut** — An underground passage connecting two parts of the mine.

**Fig. 22**

depositing the minerals, it is likely that most of the major local minerals were deposited before the end of the Carboniferous period. We do not have conclusive proof for this, and some similar vein deposits are known to occur in rocks of Permo/Trias age elsewhere in Northern England. A more complete understanding of the effect of temperature and pressure on the mineral-bearing capacity of rising brine solutions is very important for future mineral prospecting, and it could be that such a mechanism is linked with the formation of oil and natural gas deposits of similar origins.

Many mineralised faults and joints are often highly inclined to the vertical and therefore the mines formed to extract them are often very deep. It is common sense to be careful when examining specimens at disused mine workings (remember a fence erected around a hole is to keep you out, not in!). Vein ores usually carry non-metallic minerals along with them. These were deposited at the same time as the ore and are usually called **Gangue** minerals, e.g. calcite, barytes, fluorspar.

Certain areas in Cumbria have supposedly been mined since Roman times. Evidence of Roman activity is often speculative, and it is not until the Monastic influence in Mediaeval times that we have records of mining activity. The peaks of the Cumbrian metal mining industry occurred between Elizabeth I's time and the mid 1800's, when, due to technical mining problems, exhaustion of high level deposits and cheap foreign imports, the Cumbria lead/copper/zinc industry virtually finished. Mining did carry on however for a small number of specialist minerals — e.g. barytes, fluorspar, wolfram — and the two world wars gave temporary impetus to the re-opening of certain mines for a short while. The present rise in prices on the mineral market has resulted in the recent re-opening of an old mine near Keswick where there are thought to be proven reserves of galena, sphalerite and barytes to last at least for some time. As the whole history and development of Cumbria metal-mining is very well documented, no further details will be given here [21].

Certain areas in Cumbria contain concentrations of mineralised faults and fissures which carry a variety of deposits — Coniston, Threlkeld, Newlands, Catbells, Coledale, Thornthwaite, Ullswater, Dufton and Caldbeck. Fig. 22 shows the most typical mining techniques employed in the extraction of vein deposits, and cross-sections of workings can be found in most books on the subject.

(b) **Replacement Deposits :** These include the iron ore deposits of the Furness district. This type of ore deposit has a basic difference from the vein deposit; it was formed from colder mineralised solutions which came downwards (i.e. vein deposits originated from mineralised soluions which came upwards). The iron ore most commonly associated with the Furness deposits is an oxide called **Hematite** (sometimes spelt Haematite). It occurs in a variety of ways, perhaps the most attractive being the "kidney ore" which is synonymous with the

(21) **Refs**. Postlethwaite; Shaw.

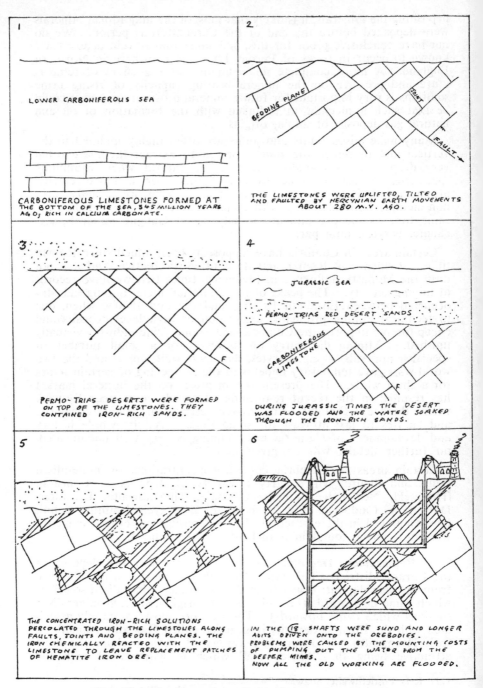

Fig. 23: The Formation of the Furness iron deposits.

area [22]. Diagram 23 explains the formation of these deposits. Tracing through geological history we start with the Carboniferous Limestones which were deposited about 345 million years ago. This great thickness of almost pure calcium carbonate was uplifted and tilted by the Hercynian earth movements which marked the end of the Carboniferous period about 280 million years ago. During the Permian and Triassic periods which followed the whole area was covered by a great thickness of desert sand. The action of the sun and lack of moisture caused the sands to become very rich in iron oxides; hence the red colour of the Penrith Sandstones, for example.

At the end of the Triassic times about 195 million years ago, the desert was flooded by the Jurassic Sea and much of the iron was dissolved by the water. Gradually the iron rich solutions soaked into the old desert sands and percolated downwards until they reached the inclined Carboniferous Limestones below. The iron-rich solutions were able to gain relatively easy access into the limestones along the dipping bedding planes and joints which these contained. By a slow process of chemical reaction, which could also involve an additional enrichment from rising brine solutions, it was possible to replace large patches of limestone with iron oxide, especially where faults channelled their flow [23]. This resulted in the formation of ore bodies along the dip of the strata. In some areas these deposits are called "churns". The random formation of such churns along the bedding plane made it difficult for the old miners to predict accurately where the most valuable deposits would be. Although iron mining has almost ceased in Furness, it is suspected that there is quite a substantial amount of iron ore still awaiting exploration when the market is right. Iron ore is not, however, one of the rarest of the world's minerals and at the moment adequate supplies can be obtained more economically from other sources.

(iii) **Evaporite Deposits** : These describe the gypsum and anhydrite beds in the Vale of Eden, and their origin ties in nicely with that of the Furness iron deposits. Diagram 24 shows that as the Permian period approached the Carboniferous Sea was subjected to increased heat. This caused the sea slowly to evaporate and as it did so the minerals which were dissolved in the water were deposited in layers on the sea bed. This resulted in the gypsum deposits which are worked by the British Gypsum Co. at Kirkby Thore between Penrith and Appleby [24]. Similarly, the sea which evaporated in the Avon

(22) The Beckermet Mine is now the only working iron mine in Cumbria. Currently they are working a large ore body measuring about 100 yards wide, 100 yards high, and 1 mile long. The workings extend to about 2,000 feet below the surface.

(23) The iron deposits are often associated with other minerals—quartz, barytes, calcite, aragonite, dolomite, specularite, and sometimes fluorspar in small amounts.

(24) Ref. Arthurton R.S. (1971).

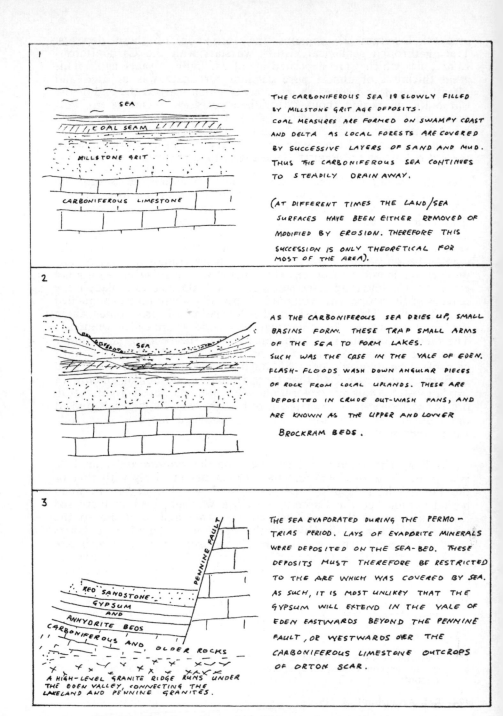

**1**

SEA

COAL SEAM

MILLSTONE GRIT

CARBONIFEROUS LIMESTONE

THE CARBONIFEROUS SEA IS SLOWLY FILLED BY MILLSTONE GRIT AGE DEPOSITS.
COAL MEASURES ARE FORMED ON SWAMPY COAST AND DELTA AS LOCAL FORESTS ARE COVERED BY SUCCESSIVE LAYERS OF SAND AND MUD. THUS THE CARBONIFEROUS SEA CONTINUES TO STEADILY DRAIN AWAY.

(AT DIFFERENT TIMES THE LAND/SEA SURFACES HAVE BEEN EITHER REMOVED OF MODIFIED BY EROSION. THEREFORE THIS SUCCESSION IS ONLY THEORETICAL FOR MOST OF THE AREA).

**2**

SEA

AS THE CARBONIFEROUS SEA DRIES UP, SMALL BASINS FORM. THESE TRAP SMALL ARMS OF THE SEA TO FORM LAKES.
SUCH WAS THE CASE IN THE VALE OF EDEN. FLASH-FLOODS WASH DOWN ANGULAR PIECES OF ROCK FROM LOCAL UPLANDS. THESE ARE DEPOSITED IN CRUDE OUT-WASH FANS, AND ARE KNOWN AS THE UPPER AND LOWER BROCKRAM BEDS.

**3**

PENNINE FAULT

RED SANDSTONE
GYPSUM AND ANHYDRITE BEDS
CARBONIFEROUS AND OLDER ROCKS

A HIGH-LEVEL GRANITE RIDGE RUNS UNDER THE EDEN VALLEY, CONNECTING THE LAKELAND AND PENNINE GRANITES.

THE SEA EVAPORATED DURING THE PERMO-TRIAS PERIOD. LAYS OF EVAPORITE MINERALS WERE DEPOSITED ON THE SEA-BED. THESE DEPOSITS MUST THEREFORE BE RESTRICTED TO THE ARE WHICH WAS COVERED BY SEA. AS SUCH, IT IS MOST UNLIKELY THAT THE GYPSUM WILL EXTEND IN THE VALE OF EDEN EASTWARDS BEYOND THE PENNINE FAULT, OR WESTWARDS OVER THE CARBONIFEROUS LIMESTONE OUTCROPS OF ORTON SCAR.

**Fig. 24 : The Formation of the Eden Valley gypsum deposits.**

district in S.W. England deposited gypsum and a rare strontium mineral — celestine. Fossil evidence indicates that the evaporating Eden Valley lake was associated with vegetation, as fossil plant remains have been found in sandstones near Hilton. This was also the dinosaur age, but little is known of any which may have roamed these parts at that time — although it could be that fossil traces of dinosaurs are awaiting the keen explorer!

Diagram 24 shows the sequence of the deposit formation. Rocks and minerals of this type are by no means restricted to this area of Cumbria alone; similar outcrops occur on the coast between Barrow and Whitehaven.

(iv) **The Diatomaceous Earth** deposits which were recently worked near Kentmere represent a deposit of organic material which originated in a glacial lake. This is quite unique in Cumbria and, although it has some affinities with the gypsum deposits (i.e. it is associated with a lake), the diatoms were minute animals which naturally accumulated on the lake-bed, whereas the gypsum was a chemical deposited on the bed of an evaporating lake. There is also a discrepancy of over 200 million years between their respective formations.

(v) **The Coal Measures :** These deposits give name to the **Carboniferous** period — which means coal-forming period. Coal is a fossil fuel formed from vegetation growing on a delta which covered Cumbria about 300 million years ago. Diagram 25 shows the successive developments in the formation of a coal seam. It is thought that the whole of Cumbria may once have been covered by coal but most of it has since been removed by erosion. The British coalfields were among the first to be geologically mapped in detail and a cross-section of a Coal Measure area shows how much faulting has taken place. Sometimes the coalminer would "lose" a whole seam as it was displaced several hundred feet up, down or sideways by vast earth movements. Coal Measure fossils can often be collected from pit-heaps. Sandstone, shale, ironstone and fire-clay are all rock types associated with coal seams. The sea coal found along the Cumbrian coast represents eroded seams which are in the process of sedimentation. Several Whitehaven collieries extend under the sea-bed.


## 15. Are there any interesting fossils to be found in local rocks?

Fossils are remains of former living creatures, which have been preserved in some way. They are of great value to the geologist in correlating the age of rocks in different parts of the world. The whole aspect of the origin of life and its evolution into our present day species is beyond the compass of this work.

Fossils only usually occur in sedimentary rocks. Igneous rocks, being formed from molten material, are very unlikely to contain fossils; likewise metamorphosed sedimentary rocks probably have had

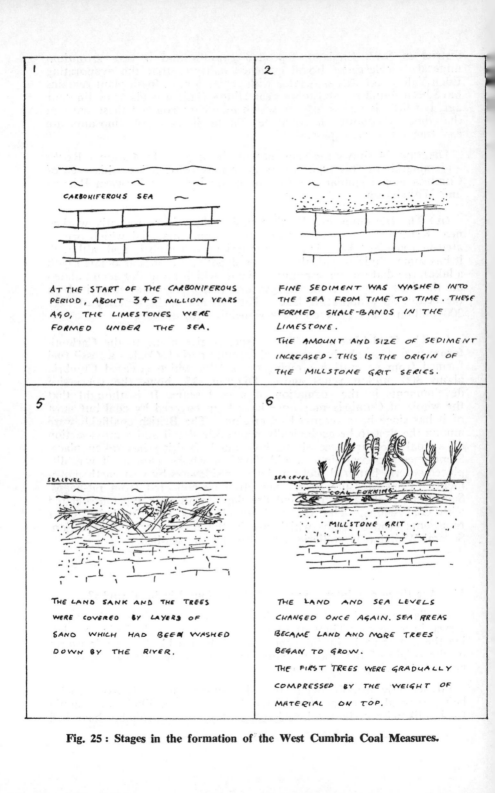

**1**

CARBONIFEROUS SEA

AT THE START OF THE CARBONIFEROUS PERIOD, ABOUT 345 MILLION YEARS AGO, THE LIMESTONES WERE FORMED UNDER THE SEA.

**2**

FINE SEDIMENT WAS WASHED INTO THE SEA FROM TIME TO TIME. THESE FORMED SHALE-BANDS IN THE LIMESTONE.
THE AMOUNT AND SIZE OF SEDIMENT INCREASED. THIS IS THE ORIGIN OF THE MILLSTONE GRIT SERIES.

**5**

SEA LEVEL

THE LAND SANK AND THE TREES WERE COVERED BY LAYERS OF SAND WHICH HAD BEEN WASHED DOWN BY THE RIVER.

**6**

SEA LEVEL

COAL FORMING

MILLSTONE GRIT

THE LAND AND SEA LEVELS CHANGED ONCE AGAIN. SEA AREAS BECAME LAND AND MORE TREES BEGAN TO GROW.
THE FIRST TREES WERE GRADUALLY COMPRESSED BY THE WEIGHT OF MATERIAL ON TOP.

**Fig. 25 : Stages in the formation of the West Cumbria Coal Measures.**

**3**

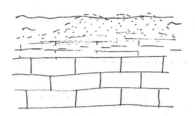

A DELTA WAS FORMED AT THE MOUTH OF A RIVER WHICH DRAINED FROM A LANDMASS TO THE NORTH. THIS DELTA COVERED MUCH OF NORTHERN ENGLAND.

**4**

VEGETATION BEGAN TO GROW ON THIS SWAMPY, COASTAL DELTA.

**7**

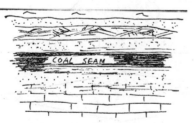

COAL SEAM

THE LAND SANK AGAIN AND THE TREES WERE COVERED BY MORE SEDIMENT.
THE FIRST TREES WERE NOW SLOWLY TURNING INTO A COAL SEAM.

**8**

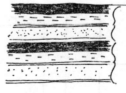

COAL
FIRECLAY (GANISTER)
SHALE
SANDSTONE
SHALE
COAL
FIRECLAY
SHALE
SANDSTONE
SHALE

THEREFORE EACH SEAM OF COAL CAN BE SEEN TO REPRESENT ONE CYCLE OF UPLIFT AND SINKING. ROCKS AND STRATA OFTEN TEND TO FOLLOW SUCH A RYTHMIC PATTERN.

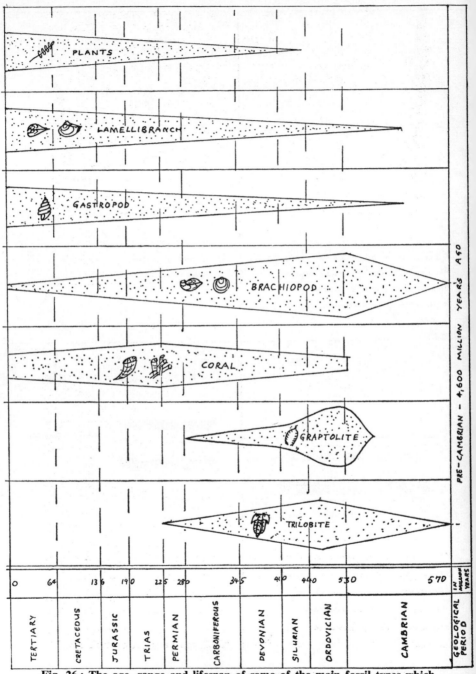

Fig. 26 : The age, range and lifespan of some of the main fossil types which have been found in Cumbria.

Remains of old cottages near Penrith which were built into the local Red Sandstone.

Typical crystals of Calcite ($CaCO_3$) formed by chemical action and water on limestone near Grange-over-Sands.

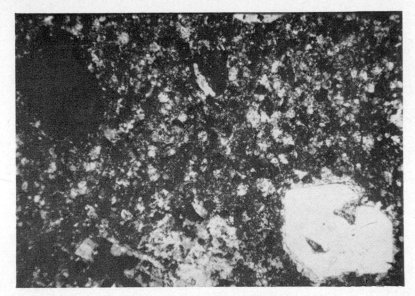

Microscopic view of Armboth Dyke Microgranite showing a dark rectangular feldspar crystal (top left) and a lighter coloured hexagonal quartz crystal (bottom right) in a groundmass of smaller crystals.

Left : Fossil coral from the Carboniferous Limestone rocks near Shap. The corals have been gradually etched out in the peaty soil. Right : Fossil coral (Lithostrotion) from the Carboniferous Limestone rocks near Cockermouth.

Piece of hematite iron ore from the Egremont district of Furness. This variety is called Kidney Ore and is used for making jewellery.

Microscopic view of microfossil found in a piece of Carboniferous Limestone rock from Orton Scar. These fossils are called Foraminifera.

| GEOLOGICAL PERIOD | | AGE AGO IN MILLIONS OF YEARS | SEDIMENTARY ROCK TYPE | FOSSILS TO BE SEEN | LOCATION |
|---|---|---|---|---|---|
| RECENT | | 2 | CLAY | DIATOMS | KENTMERE |
| JURASSIC | | 190 | JURASSIC AND LATER AGE ROCKS ARE EITHER ABSENT OR NOT EXPOSED IN CUMBRIA. | | |
| PERMO-TRIAS | | 280 | SANDSTONE | PLANT REMAINS. FOSSILS ARE RARE. | HILTON NEAR APPLEBY |
| CARBONIFEROUS PERIOD | COAL MEASURES | | COAL SHALE SANDSTONE | TREE ROOTS, BARK, SEEDS, LEAVES, SEA SHELLS | WHITEHAVEN AREA |
| | MILLSTONE GRIT | | GRITSTONE SANDSTONE | FOSSILS ARE RARE. SOME GONIATITES | MALLERSTANG. |
| | LIMESTONE | 345 | LIMESTONE MUDSTONE SHALES | CORALS, BRACHIOPODS, CRINOIDS, | ORTON |
| SILURIAN | | 440 | SHALES MUDSTONE GREYWACKE | BRACHIOPODS | AMBLESIDE |
| | | | LIMESTONE SHALE | TRILOBITES | |
| OROVICIAN | | 530 | | GRAPTOLITES | LONG SLEDDALE DRYGILL CONISTON |

Fig. 27 : **Fossiliferous rocks and localities for some different ages in Cumbria.**

all trace of original fossils either totally removed or grossly distorted. Therefore we will confine our attention to the Cumbria sedimentary rocks. Two categories emerge — water dwellers and land dwellers. Rather than consider these separately, it is possibly more helpful to consider both in the order in which they occur on the Geological Column. This is illustrated in Diagram 27. Not all the species have adapted to the changes in environment which our area has undergone in the past. Graptolites and trilobites are now extinct. Fig. 26 shows the stratigraphical life span of the fossils which you are likely to see in the Lake District rocks. If you decide to make your own collection of local geological specimens you will find it quite easy to develop a good collection of rock types as rock is very abundant. It is less easy to build up a good collection of individual minerals because there are not as many samples to be found. It is almost impossible (and undesirable) to collect a full range of local fossils as many of these are quite rare and not too well preserved — be warned and take heart [25].

## 16. What was the most recent geological activity in the Lake District?

The obvious answer to this is the **glaciers** of the last Ice Age; but this is not quite true. The whole content of this book is an attempt to show the dynamic nature of change and geological process. By this token the most recent geological process is whatever is happening as you read this — it could be a sheep causing a scree-run at Wastwater, an idiot rolling a boulder off Striding Edge, a mechanical digger excavating for road improvement near Workington, the endless capacity of running water to erode and deposit, etc. Therefore the present is just a stage in the geological evolution of the Earth, but enough of such philosophical considerations. It will suffice to say that, in so far as the "present is the key to the past", we look closely at current patterns and processes and then examine rocks to see if similar events are indicated there (the old 18th Century "Testimony of the Rocks", "Record in the Rocks", idea).

Returning to the effects of **glaciation,** the present landscape of Cumbria has been greatly influenced by ice action within the last million years. There are several distinctive features attributable to glaciation, two of which are:

(i) The result of **ice erosion** is best seen in the central mountainous part of Lakeland and is typified by U-shaped valleys, truncated spurs, hanging valleys, corrie lakes, etc.

(ii) The result of glacial **deposition** is found around the edges of the central mountains. Just as running water erodes, transports and deposits, so does a glacier. The rock removed by ice action is carried slowly along in or on the ice as it moves downhill [26].

(25) See the Geologists' Conservation Code published by the Nature Conservancy.

(26) Evidence from near Appleby suggests that the back-pressure of the glacier may have caused the ice and subglacial melt-water to flow uphill.

**Distribution of some glacial erratics**

**Fig. 28:** The dots show the locations of glacial erratics of supposed Lake District origin. These are pieces of rock which have been moved by the glacier. When the glacier started to melt, the boulders were dropped wherever they happened to be.

By plotting the occurrence of such erratics it is possible to build a map of possible ice movements during the last Ice Age. Erratics usually consist ;of large rounded blocks which would probably have been too heavy to be transported by our present rivers and streams.

Smaller pieces of rocks were deposited over a wide area by the outwash streams resulting from the melting ice. This "superficial" deposit which covers much of the North of England is called **Boulder Clay** and often constitutes a large part of the Geological Survey 1" **Drift** Edition Maps.

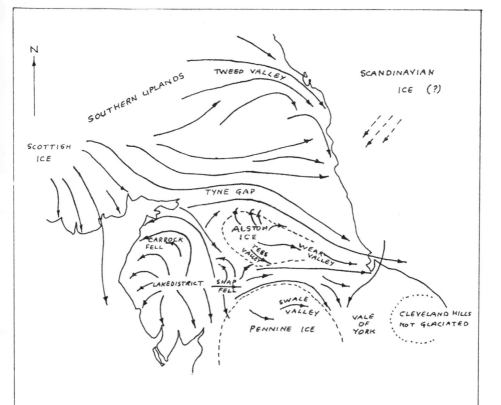

**Supposed directions of glacial movements during the last Ice Age**

**Fig 29:** Using the available evidence—drumlins, erratics, glacial drainage channels, moraines, etc., the above map has been constructed. There is a reasonable amount of agreement on the direction of flow of major glaciers, but it is still uncertain what happened where several different ice flows met.

The main Lake District valleys seem to have been affected by ice which was responsible for giving the characteristic "U" shaped valleys and lakes. In areas between the Lake District proper and the Vale of Eden, there is still confusion as to the work of the local ice in the Shap, Whinfell and Howgill areas. A careful plotting of small scale breaks in slope surfaces might help explain the patterns of ice flow. Notice the limits of Lake District, Shap and Howgill erratics. This indicates the outer limits reached by the glacier.

## Figure 30: The Stratigraphy of Rocks in Cumbria

**Superficial deposits** — often obscure solid formations. These are shown on the "Drift" editions of Geological Survey Maps.
**Recent** — Peat, river deposits and terraces.
**Glacial** — Gravels, laminated clays, moraines, boulder clay.
Solid formations underlie the superficial covering and are shown on the "solid" Geological Maps.

## PERMO-TRIAS (New Red Sandstone)

**Stanwix Shales:** Red and green shales and clays.
**Kirklington Sandstone:** Red with rounded quartz grains.
**St. Bees Sandstone:** Red-brown micaceous sandstone with shale bands.
**St. Bees Shale:** Red shales and mudstones with gypsum.
**Penrith Sandstone:** Coarse-grained red sandstone.
Great unconformity

## CARBONIFEROUS

**Upper Carboniferous**
  **Upper Coal Measures:** Red standstone and shales.
  **Middle and Lower Coal Measures:** Sandstones, clays, shales with coal seams; sometimes reddened by hematite staining.
Non-sequence
  **Millstone Grit Series:** Sandstones and grits with alternating limestone and shale; few thin coal seams. Sometimes reddened.

**Lower Carboniferous**
  **Limestone Series:** Chief Limestone Group; limestones with few sandstones, shales and thin coals.
  **Cockermouth:** Basalt lavas.
  **Basalt Conglomerate:** Conglomerate with shale bands.
Caledonian Unconformity

## SILURIAN

**Kirkby Moor Flags:** Fossiliferous greywacke.
**Bannisdale Slates:** Mudstone and silts.
**Coniston Grits:** Grey-green greywacke.
**Coldwell Beds:** Fossiliferous siltstones and grits.
**Brathay Flags:** Grey-blue mudstones.
**Stockdale Shales**
  **Browgill Beds:** Mudstone with graptolites.
  **Skelgill Beds:** Black fine-grained graptolitic shales.

## ORDOVICIAN

**Coniston Limestone Series**
  **Ashgill Beds:** Blue shale, mudstones, limestones — very fossiliferous.
  **Stockdale Rhyolites:** Pink coloured.
  **Stile End Beds/Dry Gill Shales:** Fossiliferous coarse to fine grained sediments.
Unconformity

**Borrowdale Volcanic Group**
  **High Ireby Group:** Andesitic lava, rhyolite, ashes and tuffs.
  **Binsey Group:** Andesitic lava, ashes, tuffs and agglomerates.
  **Skiddaw Slates :** Slates, sandstones, grits, shales, mudstones.

**Intrusive Igneous Rocks**
  **Skiddaw Granite:** Intruded into Skiddaw slates in Devonian times.
  **Carrock Fell Complex :** Acid and basis igneous rocks.
**Minor Intrusions:** Numerous dykes; basic and acid rocks of various composition.

**Metamorphic Rocks**
  **Skiddaw Granite Aureole:** Hornfels inner zone, spotted and chiastolite outer zone.

Much of the rock in the glacier has been ground down into clay and is washed out and deposited as **Boulder Clay.** The boulders which have not been ground down are left by the retreating glacier where the ice drops them, or where the melt-water carries them to. In many cases these boulders have been moved a long way from source. They are called **glacial erratics.** By plotting their locations it is possible to track the movements of glaciers, ice sheets and outwash of the latest glaciation. Results are most conclusive when a distinctive rock such as Shap Granite is used as a tracer. Beware however, many of the large boulders you may see as you drive along Cumbrian roads are not **glacial** erratics, but have been dumped there by farmers reclaiming land for ploughing or by road improvement schemes (human erratics!). Our area has been subjected to at least three separate glaciations and the full sequence of activities is not yet fully understood. Diagrams 28 and 29 show the conjectural directions of ice flow based on evidence as outlined previously.

## 17. What are the pit-falls if I want to make a more serious study of the Geology of Cumbria?

There is no single answer to this :

(i) People often misuse geological guide-books as "destruction guides". If you visit a published exposure and find it "worked out", look around the immediate area for another exposure. It is possible that the book's description holds good for more than one precise location, however.

(ii) When first visiting a site, allow yourself at least quarter of an hour to "get your eye in". Collecting good specimens is a long job. Look around for a little pile of specimens which has been discarded by a previous visitor. At least this will show you some of the things which interested somebody else of like mind!

(iii) If you overcollect, **do not** throw away your unwanted specimens where they will confuse somebody else. Enough confusion already results from the Forestry Commission laying paths with zinc-rich hardcore.

(iv) It is very difficult to learn local geology from books. Rocks and minerals seldom resemble the idyllic book illustrations and descriptions. It is best to "latch onto" somebody who has a first hand knowledge, or pay regular visits to compare your specimens with the labelled regiments at Carlisle and Keswick Museums.

(v) Don't look at your geological collection in terms of cash; it debases a story going back at least 4,600,000,000 years! Above all, geology is a simple physical science which can offer a lifetime of interest if you wish to know more about Cumbria than you read in the local paper.

# LOCAL AND NATIONAL STRATIGRAPHY

The following maps show the varied distribution between land and sea during the periods of geological time from Pre-Cambrian to the present times.

**Name of Period:** PRE-CAMBRIAN ERA.   **Duration:** 600 - 4,600 my.

There are no known deposits of rocks of this age in the Lake District.

The rocks at Ingleton, North Yorkshire, were thought to be Pre-Cambrian, but there is some doubt about this now. The feeling is that they are of Cambrian age.

The Pre-Cambrian Era has now been sub-divided into systems. Fossil remains are not common and tend to be very poorly preserved. Two major Earth Movements altered rocks in Scotland about 1500 and 2500 my ago.

Further details on this Era can be found in most geology textbooks.

**Name of Period:** CAMBRIAN.     **Duration:** 70my.     **Time ago:** 570my.

**Max. Sed.:** 40,000ft.

NATIONAL                    LOCAL

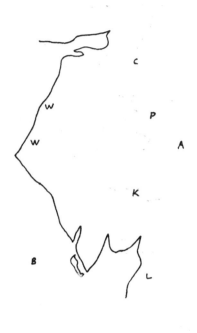

Wide geosyncline with spreading floor. Slow elevation later.

Cold to moderate climate.

Life: Age of algae and invertebrates. Dominance of trilobites, start of graptolites. Slow increase in most forms of molluscs and corals after a rather sudden appearance.

No known igneous activity.

Doubtful outcrops of Upper Cambrian only. These are mudstones and silts formed in the geosyncline. They mainly outcrop in the Isle of Man.

Skiddaw Slates could have their lowest beds of Cambrian age. The lowest beds contain a series of sediments; conglomerates, grits, flags, greywacke, shale, slates and limestones.

Rocks are zoned by trilobites.

**Name of Period:** ORDOVICIAN.　　**Duration:** 70my.　　**Time ago:** 500my.
　　　　　　　　　　　　　**Max. Sed.:** 40,000ft.

NATIONAL　　　　　　　　　　　　　　　　LOCAL

BORROWDALE
VOLCANIC
GROUP

SKIDDAW
SLATE

| NATIONAL | LOCAL |
|---|---|
| Minor earth movements and erosion. Sediments deposited in the Caledonian Geosyncline. | Skiddaw Slates pass from grits to flags and greywacke. Graptolitic shales and interbedded lava and ashes. Loweswater Flags and Cross Fell shales of this age. Unconformity and erosion. Followed by the Borrowdale Volcanic Group. |
| Folding, uplift and erosion. | These could be from submarine volcanoes (?). Fine tuffs at Sty Head, Great Langdale, and Esk Hause show current bedding. BUT there isn't much pillow-lava, or any fossils. (Compare with Snowdonia.) Well bedded, fine grained tuffs of Honister and Broughton. Over 4,000ft. of BVG. |
| Main period of Welsh Volcanics. | |
| Warm climate with first reptiles. Increase in trilobites, graptolites. molluscs and corals. | Coniston limestone stretches from Kentmere to Furness — limestones, conglomerate, ashy sandstones; contain brachiopods. graptolites. esp. in Dufton Shales. Carrock Fell and Dry Gill mudstones — trilobites. |
| | Stockdale Rhyolites interbedded. Cautley Volcanics — rhyolitic tuff and lava. |
| Spasmodic volcanics. | Ashgill Limestones and Shales. |
| | Haweswater dolerite dyke. |
| Water becomes shallower. | (Note: Yellow tuffs — rhyolitic, Green tuffs — andesitic.) |

**Name of Period:** SILURIAN.     **Duration:** 40my.     **Time ago:** 436my.

**Max. Sed.:** 70,000ft.

NATIONAL                LOCAL

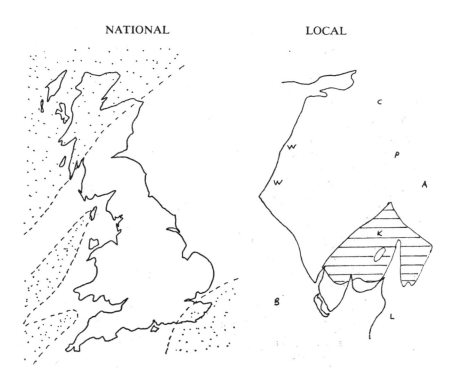

Conformable with Ordovician.

Geosyncline narrows and starts to silt up.

Caledonian movements start.

Minimal igneous activity.

Warm climate, lagoon conditions develop.

Decrease in graptolite, but increase in corals, crinoids, brachiopods and jawless fishes.

Slow infilling of the basin by SE flowing currents.

Main outcrops in three areas: S. Lakeland Fells/Kendal.; N. Furness; Howgill Fells. Fold axis NE/E-W.

Cross Fell Inlier rocks also of this age.

Stockdale Shales, Kentmere, Skelgill are outcrop areas of graptolitic mudstones. Fossils are sometimes coated by pyrite. Shales are very carbonaceous — 3.7% Carbon and 2% sulphur.

Coniston Grits.

Bannisdale Slates, series of mud and siltstones about 1,540ft. thick. Shows turbidity.

**Kirkby Moor Flags** — greywacke and grits washed down from the NW, banded.

Scout Hill Flags — grey/red current-bedded siltstones and sandstones.

**Max. Sed.:** 62,800ft.

NATIONAL                                           LOCAL

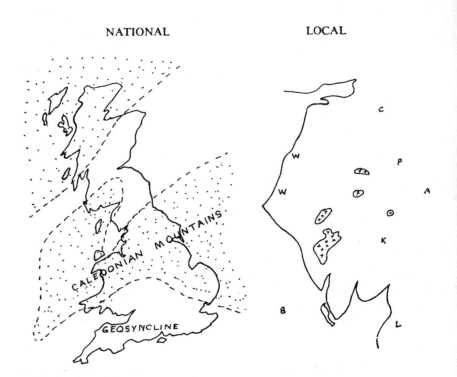

| | |
|---|---|
| Caledonian Earth movements come to end. Scotland, Lake District and Wales all uplifted. | Carrock Fell complex of gabbro and granophyre before the Skiddaw Granites (could be Ordovician). |
| Associated igneous activity. | |
| Crustal compression from NW to SE. Strata buckled, caused slaty cleavage. | Tebay folds and faults. |
| Climatic change to desert and sea in land-locked basin. | Dufton quartz-feldspar-mica porphyry into the Dufton Shales. |
| Amphibians evolve from air-breathing fishes. | |
| First insects and spiders. | Shap, Skiddaw, Eskdale granites; Ennerdale granophyre; and Carrock Fell gabbro all well exposed. |
| Last graptolites. | |
| The age of fishes. | |
| Desert rocks called OLD RED SANDSTONE. | |
| Marine deposits only in SW England. | Weardale, Askrigg and Wensleydale granites exposed about this time. |

**Name of Period:** CARBONIFEROUS (i).  **Duration:** 20my.  **Time ago:** 346my.
**Max. Sed.:** 75,000ft. (Total)

This only deals with the LOWER CARBONIFEROUS PERIOD — limestones.

### NATIONAL                               LOCAL

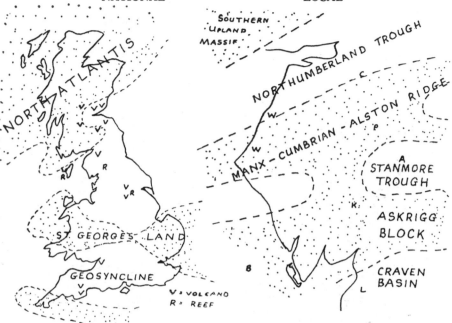

| General sinking and marine transgression from the south. | Conglomerates, screes, boulder-beds. |

General sinking and marine transgression from the south.

Caledonian Axis from Isle of Man to N. Pennines divided the seas to N and S.

Volcanic eruptions.

Basal conglomerates widespread.

Intrusion of Cornish granites and Northern dyke swarms.

Warm climate, settled conditions with increase in amphibians, reptiles and insects.

Carb. rocks contain most of our economic minerals — coal and vein ores.

Conglomerates, screes, boulder-beds.

Mell Fell north of Ullswater — 1,500 ft. of red conglomerates and coarse sandstones. Ravenstonedale — shales, sandstones and dolomitic limestones.

Shap Wells — red, felspathic conglomerate.

W. Cumb. — Cockermouth olivine-basalts spread over thin basal conglomerate, 100ft.

Furness — 90-240 metres of shales, sandstones, conglomerates and thin limestones.

Shap / Appleby — Orton dolomitic limestones and Brownber pebble bed.

Widespread outcrops of well bedded and jointed massive limestones with thin shale bands (springs).

Carb. limestone is the most cavernous of all British limestones (Ingleton).

Many fossils esp. brachiopods, corals.

Hard water and lack of surface drainage.

**Name of Period:** CARBONIFEROUS (ii).   **Duration:** 45my.   **Time ago:** 300my.

This only deals with the MIDDLE AND UPPER CARBONIFEROUS —
Millstone Grit and Coal Measures.

| NATIONAL | LOCAL |
|---|---|

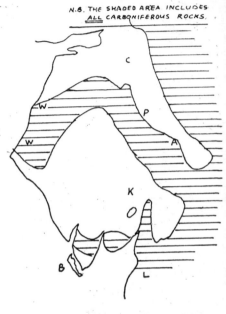

N.B. THE SHADED AREA INCLUDES
ALL CARBONIFEROUS ROCKS.

Start of Hercynian Earth Movements.

Elevation and faulting of Britain.

Scottish and Derbyshire lava flows.

Detrital material washes south from northern landmass to form large delta.

Swamp vegetation thrived.

Gently, fluctuating land levels.

Much iron from northern deserts locked into shales — red.   Mainly siderite but some pyrite.

Few fossils in grits.

Coal measure plant fossils thrived in Florida-type conditions.

Major folding starts to intensify.

Little or no sediments washed to W Cumb. or Vale of Eden because o Manx-Cumbrian Ridge. W. Cumb later got Hensingham Grit which outcrops at Whitehaven and Cockermouth. Limestones move upwards into rhythmic sequence of Yoredale sediments — Mallerstang. Workington and Whitehaven Coalfields. S. Cumb. and Furness — 275 metres of black shale, thin sandstones and limestones.

Clay band ironstones and pyrite coated fossils.

Non-marine lamellibranches, goniatites and plants associated with shale bands and coal seams.

Ganister and fireclay as seat earth below coal seam.

**Name of Period:** PERMIAN.     **Duration:** 55my.     **Time ago:** 281my.

**Max. Sed.:** 19,000ft.

NATIONAL                    LOCAL

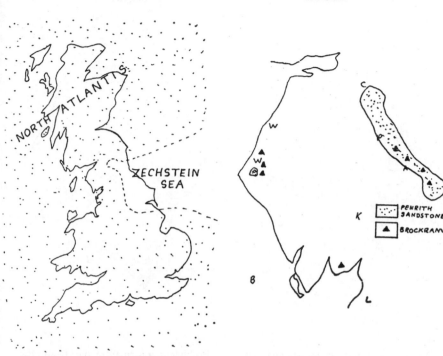

End of Hercynian earth movements. Rapid denudation of mountains.

Swamps become more arid — NEW RED SANDSTONE.

N/S ridge of the Pennines is faulted. Upfolding protects some coalfields in synclines.

Second continental phase starts.

Igneous intrusions.

Desert weathering produces much wind-blown sand.

High temperature and low humidity cause lakes to evaporate.

Only mat-algae able to survive. Last trilobites.

Evaporite deposits formed as lakes dry up.

Rocks follow conformably from Carb. Period.

Breccia and Brockram — exposed near Appleby.

W. Cumb. coalfields protected.

Whin Sill intruded — 5,000 sq. miles area.

Millet seed grains of wind-blown Penrith Sandstone.

Magnesium Limestone E. of the Pennines.

Gypsum and anhydrite deposits in Eden Vale.

### NATIONAL

### LOCAL

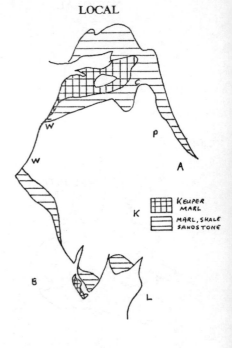

Deposition of red sandstones and red/green marls in shallow "Dead" Seas.

High evaporation continues in red desert.

Triassic sands could have covered all the Pennines — but since eroded.

Much current-bedded, wind-blown dust.

Acme of ammonites.

First dinosaurs and large marine reptiles.

First small mammals.

Britain eroded to a flat, low-lying, arid desert.

Follows conformably on from the Permian rocks, but not an obvious junction.

Keuper marls in Furness.

Furness — brockrams, anhydrite, dolomite, red muds and silts — St. Bees Shales.

Small halite deposits on Walney Island once supported small chemical industry.

W. Cumb. coast — St. Bees evaporites and sandstones.

Carlisle — basin of aeolian sandstones and evaporites.

Vale of Eden — Brockram, Eden shales and red sandstones.

Because of the similarity of the Permian and Triassic deposits, it is usual to consider them together as the Permo-Trias.

**Name of Period:** JURASSIC.      **Duration:** 57my.      **Time ago:** 180my.

**Max. Sed.:** 44,000ft.

### NATIONAL                 LOCAL

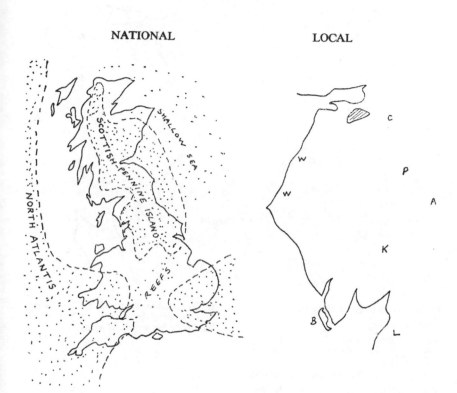

Submergence of Triassic desert below the Rhaetic Sea.

Followed by slow elevation.

No igneous activity in Europe, but great basalt lava flows in the S. Hemisphere.

Warm and humid climate, with shallow water lagoons. Very fossiliferous — many ammonites and belemnites; spread of reptiles to land, sea and air. Dinosaurs still common. Small primative mammals.

Oolitic limestones formed, outcrop runs from N. York. Moors, via Northants., to the Cotswolds.

Lias limestone and shales with clay.

Only exposure is at Great Orton, west of Carlisle.

Possible that a thickness of Jurassic material was deposited, but removed by erosion during Tertiary times.

**Name of Period:** CRETACEOUS.    **Duration:** 71my.    **Time ago:** 137my

**Max. Sed.:** 51,000ft.

NATIONAL                              LOCAL

Great world-wide increase in sea.

Britain sinks.

Very stable period allows for pure calcareous deposits to form. This is called CHALK.

Lava flows in N. Africa and Arabia.

Warm climate. Sea bordered by low lying desert.

Flints found only in Chalk.

First real birds.

Growth of reptiles.

Extinction of ammonites, belemnites, dinosaurs and large reptiles.

It is possible that the Lake District was covered by Chalk. but if so, it has since been eroded away.

**Name of Period:** TERTIARY.    **Duration:** See below:—

This period is divided into the following (oldest at the bottom)—

|  | Duration | Time ago | Max. Sed. |
|---|---|---|---|
| Pliocene | 5my. | 7my. | 15,000ft. — Alpine Earth Movements |
| Miocene | 19my. | 26my. | 21,000ft. |
| Oligocene | 12my. | 38my. | 26,000ft. — Rise of modern animals |
| Eocene | 16my. | 54my. | 30,000ft. |
| Palaeocene | 11my. | 66my. | 12,000ft. |

There was a warm climate during the Tertiary period, with sub-tropical estuarine conditions in SE England and on the Continent. This was the age of mammals.

There are no outcrops of rocks of this period in the Lake District. The present day structure and drainage, and the Pennine scarp along the Vale of Eden were affected by Alpine uplift and folding.

The N. Pennines were gently tilted eastwards.

Igneous activity produced E-SE dykes — Armathwaite Dyke which cuts through the Permo-Trias sandstones between Renwick and Dalston. This is a tholeiite (basalt) dyke belonging to the Mull swarm

**Name of Period** QUATERNARY.   **Duration:** 2my.   **Max. Sed.** 6,000ft.

| NATIONAL | LOCAL |
|---|---|

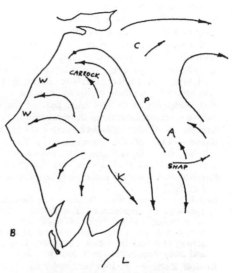

End of Alpine earth movements.

Rise of modern man.

World wide climatic changes. Could be due to tilt of Earth's axis (?).

Three or four major periods of glaciation.

Geological change is still going on, accelerated by man's activities.

Rounded hills, U-shaped valleys, ice erosion and deposition features.

Lake District and N. Pennines ice caps.

Sub-glacial melt water channel at Hilton.

Boulder clay, moraines, river terraces, raised beaches.

Peat deposits.

Langdale axe factory.

Extractive industries.

Road improvement schemes.

. . . and geological change is still going on.

# GLOSSARY OF TERMS

For terms used in the text which are not included in this glossary, please consult the Penguin Dictionary of Geology.

**Acid —**
(i) When used in conjunction with **igneous** rocks means having over 66% silica content.
(ii) When used in a chemical sense it often refers to a solution which contains an **excess of** Hydrogen ions. Hydrochloric Acid HCl is often used as a test for calcite and limestone.

**Algae** — Microscopic sea creatures.

**Amphibole** — A complex group of rock forming silicate minerals

**Andalusite** — Aluminium silicate often found in metamorphic rocks.

**Andesite** -- Fine-grained volcanic rock rich in feldspar.

**Anhydrite** — An evaporite mineral — calcium sulphate, often found with gypsum

**Anticline** — A folded structure found in rocks, a cross-section of which **appears** as an arch (opposite of Syncline, q.v.).

**Asthenosphere** — A less rigid part of the earth under the crust.

**Augite** — A rock-forming mineral rich in calcium, magnesium and iron.

**Azurite** — A blue coloured secondary ore of copper — $CaCO_3Cu(OH)_2$.

**Basalt** — One of the commonest **volcanic rocks** (q.v.). It forms extensive lava flows; the Giants Causeway in Antrim is composed of basalt. It is a dense dark-green to black rock and is very fine-grained.

**Barytes** — A gangue mineral often found in vein-ore deposits — barium sulphate $BaSO_4$

**Bedding** — A structure found in **sedimentary rocks** (q.v.) representing individual episodes of deposition. The break between beds may be a mechanical break or a break in the type of rock. Much bedding when it is formed is roughly horizontal. In some cases a **cross-bedding** (current bedding) is produced by the action of wind or water currents. This type of bedding may be relatively steep and may represent the front or side of an advancing delta, bank or dune.

**Benioff Zone** — A localised hot-spot created at the leading edge of a continental plate.

**Biotite** — A rock-forming silicate mineral which is the dark variety of mica. Has a platy crystal shape and is common in many rock types.

**Boulder Clay** — Material deposited by ice sheets and consisting of pebbles and boulders in **matrix** (q.v.) of **clay** (q.v.). The boulders can often be identified with occurrences of rock hundreds of miles away.

**Brachiopods** — A group of marine animals with shells that have existed since Cambrian times. They are common as fossils especially in **limestones** (q.v.). They resemble Roman lamps and so are sometimes called lamp shells.

**Breccia** — A rock made up of coarse **angular** fragments of rock material set in a fine-grained matrix (q.v.). Such rocks may represent old screes or are metamorphic or may represent the products of volcanic activity when they are **igneous.**

**Brockram** — A sedimentary deposit of Permian-Triassic age formed in outwash fans caused by flash floods and rapid erosion, usually of Carboniferous Limestone.

**Calcite** — A common mineral composed of calcium carbonate. It effervesces briskly with dilute (2N) hydrochloric acid giving off carbon dioxide. It is easily scratched with a knife, has a white streak, and shows good cleavages. It is the main constituent of limestones.

**Calcium** — A metal which reacts slowly with water. Doesn't occur as a metal in nature.

**Caledonian** — Major period of earth movement about 500my ago.

**Cambrian** — A geological period which lasted from 530 to 570 million years ago. From the Roman name for Wales.

**Carboniferous** — A geological period which lasted from 280 - 345 million years ago.

**Celestine** — An evaporite mineral of Triassic age which is rich in strontium.

**Cement** — The material that binds together the particles of some sedimentary rocks. The composition of cements varies but it is usually **calcite** (q.v.) **quartz** (q.v.) **haematite** (q.v.) or **clay** (q.v.).

**Cerrusite** — A secondary ore of lead — $PbCO_3$.

**Chalk** — A particularly pure form of **limestone** (q.v.) made up largely of **micro-fossils** (q.v.). It is white in colour and was typically formed during later Cretaceous times.

**Chert** — A form of silica (silicon dioxide) so fine-grained that its crystalline nature is difficult to detect. Like **quartz** (q.v.) it is harder than steel, and it is usually yellowish or brownish in colour. It occurs as discreet masses or as layers in bodies of **limestone** (q.v.) or other sediments.

**Chiasolite** — A crystalline form of andalusite found in highly metamorphosed rocks.

**Clay** — A very fine-grained sediment formed from minute flakes of aluminium rich **silicate minerals** (q.v.). Mineral particles are less than .002mm across.

**Cleavage** — Planes in some minerals along which they split readily. These planes are closely related to the atomic structure of the mineral. Some rocks, such as **slate** (q.v.) also show cleavage. These are also planes along which the rocks split easily, but they are caused by the parallel arrangement of flaky silicate minerals as a result of stress.

**Clinometer** — See **Dip.**

**Coal** — A **sedimentary rock** (q.v.) composed largely of carbon, produced by accumulation and subsequent carbonisation of vegetable matter. Most British coal is of Carboniferous age.

**Conchoidal Fracture** — The shape of fracture surfaces found in glass. A number of rocks and minerals also show this type of fracture.

**Conglomerate** — A **sedimentary rock** (q.v.) consisting of rounded pebbles or boulder set in a finer-grained matrix. (Compare with a sedimentary **breccia** in which the larger particles are angular.)

**Corals** — Common marine animals, many of which are found as fossils. The coral polyp itself is soft; it is the protective supportive skeleton (which is **out-side** the polyp when it is alive) which is preserved in rocks (mostly limestones).

**Cretaceous** — A geological period which lasted from 63 to 136my ago.

**Crinoids** — Marine fossils that resemble plants (hence their name — "sea-lilies"). They are related to the sea urchins (see **echinoids**) and starfish. Usually all that remains of them are the small discs that build up its "stem" and "branches" and which become separated from each other when the animal dies.

**Crust** — The common term used for the Lithosphere, being the outer rigid layer rocks round the outside of the Earth.

**Crystal** — Is a solid substance of uniform chemical composition bounded by well-marked planes (faces). The external shape of crystals is an expression of their atomic structure. Substances composed largely of crystals that do not necessarily show good crystal faces are said to be **crystalline.**

**Devonian** — A geological period which lasted from 345 to 410my ago.

**Dilute (2N) Hydrochloric Acid** — Is prepared by measuring off about 200ml of concentrated commercial acid and diluting it to a volume of 1 litre with distilled water.

**Diorite** — A slow cooling intermediate igneous rock.

**Dip** — Is the maximum slope on a tilting bedding (q.v.) fault (q.v.) or other structure in a rock. The amount of slope is measured in degrees from the horizontal using a **clinometer.** (See also **strike.**) In addition to recording the amount of dip in degrees it is also necessary to record the geographical direction (bearing) in which it slopes, e.g. "24° NW" or "24° towards 315°"

**Dolerite** — An intrusive igneous rock.

**Dolomite** — A mineral composed of calcium magnesium carbonate. It resembles **calcite** (q.v.) but does not effervesce with cold dilute (2N) hydrochloric acid. Rock composed largely of dolomite is called dolomite rock or dolomitic limestone.

**Drift** — (i) The superficial covering of rocks by deposits, river and glacial material. (ii) The edition of a geological map which shows the "solid" edition covered by (i).

**Dyke** — This is a relatively thin tubular body of **igneous rock** (q.v.) that cuts across the bedding of rocks it may intrude (c.f. **sill**).

**Echinoids** — Sea urchins (or "sea eggs") many of which are found as fossils.

**Epidote** — A rock forming silicate mineral, usually green.

**Erratic** — A block of rock which has been transported from its original source.

**Evaporite** — A sedimentary mineral formed as a sea dries up.

**Extrusive** — An igneous rock type formed by lava being flung out onto the Earth's surface where it cools very rapidly to form a fine-grained rock.

**Fabric** — The internal structure or arrangement of minerals in a rock. The alternative term texture is also used.

**Fault** — A fracture in rocks along which some differential movement (displacement) has taken place. Movement along a fault can crush the rocks on either side to produce a **breccia** (q.v.). Some faults are filled with a clay-like material; this is referred to as a **fault gauge.**

**Feldspar** — A group of silicate minerals containing aluminium with calcium sodium or potassium. Together they form the commonest mineral group in the Earth's crust. Different types include **orthoclase, microcline** and the **plagioclase** series.

**Felsite** — A fine-grained igneous dyke rock, often associated with plutonic masses.

**Ferro** — Rich in iron.

**Flag** — A sedimentary rock, a sandstone, which has bands of mica which allows it to be split into large thin pieces.

**Flint** — A dark grey type of chert (q.v.) found typically as modules or layers in **chalk** (q.v.).

**Fluorspar** — A gangue mineral, calcium fluoride $CaF_2$.

**Foliation** — A banding, layering, or parallel growth of minerals typically found in **metamorphic rocks** (q.v.). It resembles the "leaves" of a book.

**Fossil** — The remains of animals or plants preserved in rocks — the remains of the creature itself or simply a cast. Tracks of burrows of fossil animals are called **trace fossils.** The word fossil should only be referred to biogenic remains (i.e. those remains or traces of once living organisms).

**Gabbro** — A slow cooling, large grained igneous rock which contains very little quartz.

**Galena** — A mineral composed of lead sulphide. It is very dense, is easily scratched with a knife blade, and has a bright silvery appearance. Its streak (q.v.) is lead grey. If warmed in dilute (2N) hydrochloric acid it gives off a strong smell of rotten eggs. It has good cleavages so that it breaks into small cubes. It is the main ore of lead.

**Gangue** — The minerals which do not contain ore but are deposited in metalliferous veins.

**Garnet** — A hard, usually red **silicate mineral** (q.v.). Its crystal shape is such that it usually appears to be rounded. It is found mostly in metamorphic rocks (q.v.).

**Gastropods** — Animals with a single coiled shell; the best known living example is the garden snail. Many gastropods are marine and they frequently occur as fossils.

**Geosyncline** — A large deep trough, filled with water and often separating two plates.

**Gneiss** — A coarse-grained **metamorphic rock** (q.v.) characterised by a **foliation** (q.v.) made up of alternating bands of light and dark minerals.

**Grain** — The size of the constituent materials in a rock.

**Granite** — A coarse grained crystalline rock (see **crystals**) composed of **quartz** (q.v.), feldspar (q.v.) and a dark mineral such as biotite (see mica). Some granites are igneous rocks (q.v.) but others may be metamorphic in origin.

**Granophyre** — An igneous rock similar to granite.

**Graptolites** — An extinct group of delicate marine fossils found in Palaeozoic rocks. They appear to be little more than pencil marks made on **shale** (q.v.) or **slate** (q.v.) but are among the more important fossil groups used in dating rocks. In life they were probably planktonic.

**Greywacke** — A **sedimentary rock** (q.v.) best considered as a "dirty" sandstone with angular sandgrains of **quartz** (q.v.) and **feldspar** (q.v.) in a **matrix** (q.v.) or **calcite, clay** and other silicate minerals. It is one of the commonest types of sediment and occurs in parts of Wales, the Lake District and Scotland.

**Grit** — A variety of sandstone (q.v.) with angular quartz grains. The term is sometimes (incorrectly) used to describe any coarse sandstone and in the Cotswolds has even been applied to rubbly fossil-rich limestone.

**Gypsum** — A mineral, essentially a hydrated calcium sulphate. It is formed in some cases by evaporation of water from lakes in desert regions. It is so soft that it can be scratched with a finger nail. It has a white **streak** (q.v.). It is the main source of plaster of paris.

**Haematite** — One of the mineral forms of iron oxide (see also **magnetite).** It is greyish in colour but has a reddish **streak** (q.v.). It is an important ore of iron, often occurring in kidney-shaped masses.

**Hardness** — A property of minerals deduced from its ability to scratch or be scratched by certain substances. Common materials used for testing mineral hardness are (in increasing order of hardness): finger nail, copper coin, steel, glass, carborundum (silicon carbide). The relative hardness of two minerals may readily be determined by attempting to scratch one with another.

**Head** — A pleistocene deposit produced on the land and consisting of angular fragments of local rock in a matrix of fine material. It seems to have been formed by mass flow of material down slopes during breaks in the glaciation.

**Hercynian** — A period of great earth movements at the end of Carboniferous times about 280my ago.

**Hornblende** — A fairly common silicate mineral containing calcium, magnesium, iron and sometimes aluminium and sodium. It is a very dark green in colour and is about as hard as steel (see **hardness**). It has a white **streak** (q.v.). It is found in some **granites** (q.v.) but is also common in many **metamophic rocks** (q.v.).

**Hornfels** — A metamorphic rock formed by the action of heat and often on an existing fine-grained igneous rock.

**Igneous** — A rock that has solidified from a **magma** (q.v.). If the magma cools slowly, e.g. deep in the earth, it tends to be coarse-gained (see **granite** as an example). If the magma cools rapidly, e.g. as a lava-flow, it is fine-grained (like **basalt)** or even glassy.

**Inlier** — An area of rock on a map which is surrounded by younger rock. (See also **outlier).**

**Intrusive** — A type of igneous rock which was formed by lava cooling down in narrow cracks in the earth's crust.

**Joint** — A fracture in a rock in which there has been no differential movement (c.f. **fault**).

**Jurassic** — A geological period which existed between 136 and 190my ago.

**Lamellibranch** — Another name for a **pelecypod** (q.v.).

**Lamprophyre** — An igneous rock of a complex nature which often contains large crystals and has an intrusive origin.

**Lava** — **Magma** (q.v.) that has reached the surface of the earth; the term is also supplied to the solid rock into which lava cools.

**Lias** — The lower part of the Jurassic Period.

**Limestone** — A widespread sedimentary rock-type composed largely of **calcite** (q.v.). The calcite may be present as small crystals. as very fine grains. as fossils or fragments of fossils, or as **oolites** (q.v.). Many varieties are therefore possible (e.g. **chalk** (q.v.)). It is readily identified because it effervesces with **dilute (2N) hydrochloric acid** (q.v.).

**Limonite** — A mineral composed of iron oxide and water. It is soft, earthy and brown, yellow or red in colour. Essentially it has the same composition as rust.

**Lithology** — All the characteristics (composition, mineral content, fabric, fossil content) that make up a rock.

**Lithosphere** — The rigid outer part of the Earth's crust.

**Load** — The sediments carried by a stream or glacier.

**Magma** — Molten rock material with the consistency of porridge, composed of a mixture of melt, crystals, rock fragments and gases. (See also **igneous rocks, lava**).

**Magnesium** — A metal — Mg.

**Magnetite** — A mineral —black magnetite iron oxide. Deflects a compass needle, has a black **streak** (q.v.) and is about the same **hardness** (q.v.) as steel. It is an important ore of iron.

**Malachite** — A green coloured secondary ore of copper — $Cu_2CO_3(OH)_2$.

**Marble** — A sugary-looking **metamorphic rock** (q.v.) composed largely of calcite (q.v.). It is a metamorphosed limestone and like limestone it effervesces with **dilute (2N) hydrochloric acid** (q.v.).

**Marl** — A **mudstone** (q.v.) with an appreciable amount of **calcite** (q.v.) so that as a rock it effervesces with dilute acid. Some rocks described as marl in fact contain very little calcite.

**Matrix** — A term in describing the fine part of a rock **fabric** (q.v.) as opposed to the coarser crystals, pebbles or other fragments it may contain.

**Mechanical** — Sedimentary rocks caused by the weathering and movement of rock particles to their place of deposition.

**Metamorphic Rocks** — Those rocks that have been changed by heat, pressure or hot solutions without actually melting. The change may take the form of simple re-crystallisation (e.g. **limestone** to **marble**) or may involve complex changes in mineral composition. **Gneiss, schist,** and **slate** (q.v.) are examples of metamorphic rock.

**Meteorite** — A rock from outer space which withstands the heat of entering the Earth's atmosphere, and lands on the Earth's surface.

**Micas** — A group of flaky **silicate minerals** (q.v.) containing aluminium and usually potassium. White mica is known as **muscovite** and the commonest dark mica is **biotite.** Micas are soft; they may be scratched with fingernails. The thin flakes bend easily. but they are also elastic and spring back to shape.

**Microfossils** — Fossils only visible under a microscope such as foraminifera, radiolaria, diatoms, ostracods, spores and small fragments of larger fossils. Microfossils are extremely important in dating rocks.

**Mineral** — A naturally occurring **crystalline** substance (see **crystal**) with a definite chemical composition and uniform atomic structure.

**Molluscs** — A large phyllum of animals among which are many fossil representatives (see: **ammonites, gastropods, pelecypods**).

**Mudstone** — A sedimentary rock which is compacted **clay** (q.v.) and which breaks with a conchoidal fracture. It is soft and becomes slightly sticky when wet (like clay). Colours vary from black, grey, red, brown through to green.

**Muscovite** — One of the common forms of mica.

**Olivine** — A rock forming silicate mineral.

**Oolite** — A small (about 1mm diameter) sphere of calcium carbonate resembling a sand grain. Sections show that it is made up of concentric and radial growths of calcite. Oolites are formed in warm saline waters as a result of chemical precipitation and movement by currents. Limestones (q.v.) composed largely of **oolites are said to be oolitic** and frequently show **cross-bedding** (see bedding).

**Ordovician** — A geological period which lasted from 440 to 530 my ago.

**Organic** — Having an association with once-living things.

**Orthoclase** — One of the more common feldspars.

**Palaeontology** — The study of fossils.

**Palaeozoic** — The geological era which contains the **Cambrian, Ordovician, Silurian, Devonian** and **Carboniferous** Periods.

**Pelecypods** (also called **Bivalves** and **Lamellibranchs**) — A group of bivalve molluscs common as modern sea shells and as fossils. They may occur in **fresh** or sea water. The two valves tend to be mirror images of each other; this distinguishes them from **brachiopods** (q.v.) in which the two valves are dissimilar.

**Permian** — The geological period of time which lasted from 225 to 280my ago.

**Petrology** — The general term for the study of rocks.

**Phenocryst** —A large crystal set in a mass of smaller ones.

**Picrite** — A plutonic igneous rock, dark coloured, coarse grained and rich in olivine.

**Plagioclase** — A common variety of feldspar.

**Plate Tectonics** — A theory which suggests that there are large recognisable pieces of old rock which are floating about on the Earth's surface.

**Pluton** — A mass of igneous rock which originates deep down in the asthenosphere.

**Potassium** — A very reactive metal which immediately burst into flames on contact with water.

**Pre-Cambrian** — All the age of the Earth before the Cambrian period 570my ago. It is now sub-divided into smaller divisions.

**Provenance** — A term used to describe the source of **sedimentary** (q.v.) material.

**Pyrite** — A common, brassy coloured mineral, essentially iron sulphide in composition. It is harder than steel and often occurs in cubic shapes. It has a dark-greenish-grey **streak** (q.v.). With warm **dilute (2N) hydrochloric acid** (q.v.) it gives off a smell of rotten eggs.

**Pyroxene** — A group of rock formine silicate minerals, similar to amphiboles.

**Quartz** — A very common silicate mineral, essentially silicon dioxide. It is harder than steel, has a white **streak** (q.v.) and a **conchoidal fracture** (q.v.). It is the main constituent of sand and **sandstone** (q.v.), but is also found in **many other** rock types including **granite** (q.v.).

**Rhyolite** — A fine-grained volcanic rock (igneous).

**Rock** — An aggregate of mineral grains, geologically a rock does not have to be so hard, so that **clay** (q.v.) is as much a rock as **granite** (q.v.).

**Sandstone** — A common sedimentary rock (q.v.) composed mainly of grains of **quartz** (q.v.) ranging in size from 1/16mm to 2mm. The grains may be rounded or angular (see **grit**), and may be cemented together with a variety of minerals (see **cement**). Some sandstones show **cross-bedding** (see bedding).

**Schist** — A common group of **metamorphic rocks** (q.v.) characterised by **foliation** (q.v.) but without banding. **Micas** (q.v.) are especially common in schists.

**Scree** — A deposit of angular material that has accumulated at the foot of a cliff or other steep slope as a result of the weathering and collapse of material above (see **breccia**).

**Sedimentary Rocks** — The compacted accumulation of deposits laid down on the Earth's surface (where it is in contact with either air or water). The material may represent rock waste, chemical precipitation or accumulations of biogenic material (see **fossils**). The sediment may have been transported long distances or may occur almost where it was formed. Common sedimentary rocks include **limestones, sandstone, mudstone, conglomerate,** coal (q.v.).

**Shale** — A sedimentary rock, which is compacted **clay** (q.v.) and which splits into nne layers or laminae. Scratches easily with a knife. Does not show conchoidal fracture.

**Silicate Minerals** — A group of mineral grains characterised by a high content of silicon and oxygen. Most rock-forming minerals (except **calcite** (q.v.) are silicates (e.g. **feldspars, quartz, hornblende, micas** (q.v.).

**Sill** — This is a relatively thin tabular body of **igneous rock** (q.v.) that is essentially parallel to the bedding of rocks (c.f. **dyke**).

**Silt** — This is sedimentary material that is finer than sand but coarser than clay (i.e. from 1/16mm down to .002mm). When compacted into a rock it forms **siltstone.**

**Silurian** — A geological period which lasted from 410 - 440my ago.

**Slickensides** — Scratches produced on faults (q.v.) and other surfaces in rocks as a result of movement of one part over another. By rubbing the finger along slickensides geologists can sometimes decide in which direction the movement has taken place.

**Sodium** — A very reactive metal which melts on contact with water.

**Sodium Chloride** — The chemical composition of common salt.

**Solid** — The edition of geological map which shows the outcrops of rocks underneath the soils and other drift deposits.

**Sphalerite** — An ore or zinc, also known as Black Jack or Zinc Blende. ZnS.

**Stratum** — A single layer or bed of sedimentary rock (see bedding). Plural: Strata. The study of the history of sequences in strata is known as **Stratigraphy.**

**Streak** — A property of minerals. This is the colour of the finely powdered mineral which sometimes differs from its body colour. Streak is tested by rubbing the mineral on a piece of white unglazed porcelain or on a piece of black silicon-carbide (wet and dry) paper.

**Strike** — This is the trend of geological structures measured in a horizontal plane. In a tilted bed the strike line is at right angles to the **dip** (q.v.) and is the bearing taken along a horizontal line on the bed. Strike is usually recorded in three digits (to distinguish it from dip), e.g. 025°, 232°, 360°. 001°.

**Syncline** — A folded structure found in rocks, which resembles a trough. (Opposite of **anticline**).

**Terrestial Magnetism** — The Earth's magnetic field.

**Tertiary** — The period of geological time from 2 to 64my ago.

**Texture** — Another term for **fabric** (q.v.).

**Tholeiite** — A type of basalt.

**Trace Fossils** — See **Fossil.**

**Triassic** — A geological period which lasted from 190 to 225my ago.

**Trilobites** — An extinct group of marine anthropods (related to insects, spiders, scorpions, centipedes, lobsters) found as fossils in Palaeozoic rocks. Their bodies were segmented and protected by an exoskeleton (an outside, hard protective layer). The body is divided **lengthwise** into three distinct lobes, and there was also a head shield and sometimes a tail shield. They lived in fairly shallow water near to or on the sea floor.

**Tufa** — A cave deposit of calcite.

**Tuff** — Small grained unconsolidated material and ash thrown out of a volcano.

**Unconformity** — A marked break in the geological record preserved in rocks, brought about by a period of erosion followed by sedimentation. This results in a series of **strata** (q.v.) overlying older rocks which may have different lithologies, structure and geometrical disposition. Some unconformities can represent breaks in stratigraphy of hundreds of millions of years.

**Uniformitarianism** — One of the fundamental philosophies of geology first stated as "The present is the key to the past". This simple statement has been elaborated and extended in scope to become the **actualistic principles** of this century, which although more sophisticated than the original doctrine nevertheless preserve the same spirit.

**Vein** — Mineral infilling of a narrow fissure in a rock.

**Volcanic** — An adjective applied to all igneous material that is erupted at the surface of the Earth.

**Wolfram** — An ore of tungsten.

# SELECTED BIBLIOGRAPHY

Almond, D. C., and Whitten, D. G. A. (1976), **Rocks, Minerals and Crystals,** Hamlyn.

Arthurton, R. S., (1971), "The Permian Evaporites of the Langwathby Borehole, Cumbria", Report No 71/17 of the Natural Environmental Research Council, Institute of Geological Sciences, HMSO.

Bott, M. H. P., (1974), "The Geological Interpretation of a Gravity Survey of the English Lake District and the Vale of Eden", J. Geol. Soc., London, Vol. 130, pgs. 309-331.

Bradshaw, M. J., (1973), **A New Geology,** English Univ. Press.

Burgess, I. C., and Wadge, A. J., (1974), **The Geology of the Cross Fell Area,** HMSO.

Countryside Commission, (1975), **Lake District,** National Park Guide No. 6, HMSO.

Dixon, E. E. L., et al, (1926), **Geology of the Carlisle, Longtown and Silloth District,** HMSO.

Eastwood, T., et al, (1968), **The Geology of the country around Cockermouth and Caldbeck,** Sheet 23 Memoir, HMSO.

Fitton, J. G., and Hughes, D. J., (1970), "Earth Planet". Sci.Lett., Vol 8.

Gass, I. G., et al, (1971), **Understanding the Earth,** Open University Science Foundation Course Book, Artemis Press.

Hamilton, W. R. Woolley. A. R. Bishop, A. C. (1974), **Minerals, Rocks and Fossils,** Hamlyn.

HMSO, (1928), **Catalogues of Plans of Abandoned Mines — Vol. I.**

Institute of Geological Sciences, (1973), **The story of the Earth,** HMSO

Institute of Geological Sciences, (1974), **Bulletin of the Geological Survey,** No. 46. HMSO.

Kendall, P. F., and Wroot, H. E., (1924), **Geology of Yorkshire,** printed privately in Vienna.

Kennedy, J., (1970), **Milk Marketing Project Book, No. 130,** MMB.

King, C. A. M., (1976), **The Geomorphology of the British Isles Series: North of England,** Methuen.

Marr, J. E., (1916), **The Geology of the Lake District,** Camb. Univ. Press.

Millward. R., and Robinson, A., (1972), **Landscapes of Britain Series: Cumbria,** Macmillan.

Mitchell, G. H., (1970), **Geologists' Association Field Guide No. 2: The Lake District,** Geol. Assoc.

Monkhouse, F. J., (1972), **The English Lake District,** British Landscape through Maps Series, Geog. Assoc.

Moseley, F., (1976), "The Plate Tectonic Origins of the Lake District", Paper, Assoc. Advcmt. Sci. Congress, Lancaster 1976. Unpublished paper.

Phillips, W. E. A., Stillman, C. J., and Murphy, T. A., (1976), "Caledonian Plate Tectonic Model", Jour. Geol. Soc., London, Vol. 132. pt. 6, Nov. 1976, pgs. 579-611, Geol. Soc. London.

Postlethwaite, J., (1913), **Mines and Mining in the English Lake District,** reprint of 3rd ed. (1975), Moon, Beckermet.

Prosser, R., (1977), **Geology Explained in the Lake District.** David and Charles.

Raistrick, A., and Illingworth, J. L., (1967), **The Face of North West Yorkshire,** Dalesman.

Rayner, D. H., and Hemingway, J. E., (Eds.), (1974), **The Geology and Mineral Resources of Yorkshire,** Yorks. Geol. Soc.

Shackleton, E. H., (1975), **Geological Excursions in Lakeland,** Dalesman.

Shackleton, E. H., (1973), **Lakeland Geology,** Dalesman.

Shaw, W. T., (1972), **Mining in the Lake Counties,** Dalesman.

Smith, A. J., (1974), **Geology**, Hamlyn.

Sugisaki, R., (1976), In Lithos, Vol. 9, 17-30, **Chemical characteristics of Volcanic Rocks :** Relation to plate movements.

Taylor, B. J., et al, (1971), **British Regional Geology: Northern England,** 4th Ed., HMSO.

Waltham, A. C. (Ed.), (1974), **Limestones and Caves of North West England,** British Cave Research Assoc., David and Charles.

Warring, R. H., (1970), **Milk Marketing Board Project Book, No. 55: Be a Rock Collector,** MMB.

Whitten, D. G. A.. and Brooks, J. R. V., (1972), **Dictionary of Geology,** latest revision, Penguin.

Wright, A. E., (1976), "Alternating Subducting and the Evolution of the Atlantic Caledonidies" in Nature, Vol. 264, Nov. 1976.

Yorkshire Geological Society, numerous papers in their proceedings.

## References

This list of references has been kept deliberately short. For a more detailed lists please consult the latest editions of:—

(i) Smith, R. A., **A Bibliography of the Geology and Geomorphology of Cumbria,** Cumberland Geological Society.

(ii) Thorpe, J. A., **A Bibliography of the Geology and Physical Geography of North and West Lancashire and the Isle of Man,** University of Lancaster, Occasional Paper No. 5 (1972).

# GENERAL INDEX

(Text only, excluding direct reference to diagrams)

**J**

Jurassic : 7, 43, 67.

**K**

Kirkby Stephen : 29.
Kentmere : 45.
Keswick : 57.

**L**

Limestone : 13, 33, 39.
Lithosphere : 15.
Lava : 15.
Lamprophyre : 21.
Langdale : 23, 37.
Limonite : 37.

**M**

Metamorphic : 5, 7, 35.
Mica : 15.
Malachite : 37.
Mines : 41.

**N**

Neolithic : 37.

**O**

Ordovician : 7, 11, 13, 39, 60.
Ores : 37.

**P**

Permian : 7, 41, 43, 65.
Plate Tectonics : 11, 21.
Pluton : 19.
Penrith : 43.
Pre-Cambrian : 58.

**Q**

Quartz : 23, 43.
Quaternary : 69.

**R**

Rhyolite : 19, 23, 35.

**S**

Sedimentary : 5, 29, 33, 45.
Shap : 5, 19, 23, 35, 39, 57.
Stratigraphy : 7, 29, 58.
Silurian : 7, 21, 33, 35, 39, 61.
Subduction : 13.
Shale : 13, 33, 45.
Sca Fell : 19.
Syncline : 29.
Sandstone : 33, 45.
Silt : 33.
Sinen Gill : 35.
Skiddaw Slate : 35, 59.
Stone Axes : 37.

**T**

Triassic : 7, 41, 43, 66.
Trilobites : 13, 53, 59.
Tuff : 19.
Threlkeld : 35.
Tebay : 19, 35.
Tertiary : 68.

**V**

Vale of Eden : 19, 39, 43, 45.
Veins : 37, 41.

**W**

Wolfram : 41.
Whitehaven : 45.